# SpringerBriefs in Energy

More information about this series at http://www.springer.com/series/8903

Carlos Rubio-Bellido · Alexis Pérez-Fargallo
Jesús Pulido-Arcas

# Energy Optimization and Prediction in Office Buildings

## A Case Study of Office Building Design in Chile

 Springer

Carlos Rubio-Bellido
Higher Technical School
  of Building Engineering
Universidad de Sevilla
Seville, Sevilla
Spain

Jesús Pulido-Arcas
Faculty of Construction,
  Architecture and Design
Universidad del Bío-Bío
Concepción, VIII–Concepción
Chile

Alexis Pérez-Fargallo
Faculty of Construction,
  Architecture and Design
Universidad del Bío-Bío
Concepción, VIII–Concepción
Chile

ISSN 2191-5520          ISSN 2191-5539   (electronic)
SpringerBriefs in Energy
ISBN 978-3-319-90145-9      ISBN 978-3-319-90146-6   (eBook)
https://doi.org/10.1007/978-3-319-90146-6

Library of Congress Control Number: 2018939316

Printed on acid-free paper

This Springer imprint is published by the registered company Springer International Publishing AG part of Springer Nature
The registered company address is: Gewerbestrasse 11, 6330 Cham, Switzerland

# Contents

# Symbols and Abbreviations

| | |
|---|---|
| $A_i$ | Area of the element i (m$^2$) |
| $A_c$ | Area of the opaque surfaces (m$^2$) |
| $A_E$ | Area of the building envelope (m$^2$) |
| $A_{sol,k}$ | Effective collecting area of surface k with given orientation and tilt angle (m$^2$) |
| $A_{sol,op}$ | Effective collecting area of opaque surface k with given orientation and tilt angle (m$^2$) |
| $A_w$ | Overall projected area of the glazed element (m$^2$) |
| $b_{ve,k}$ | Adjustment factor of temperature for the airflow k |
| $C_m$ | Corrected internal heat capacity, (KJ/m$^2$K) |
| $F_F$ | Frame area fraction, ratio of the projected frame area to the overall projected area of the glazed element |
| $F_{r,k}$ | Form factor between the element and the sky |
| $F_{sh,gl}$ | Shading reduction factor for movable shading provisions |
| $F_{sh,ob,k}$ | Shading reduction factor for external obstacles for the solar effective collecting area of surface k |
| $f_{ve,t,k}$ | Time fraction of operation over the calculation period (full time: ft = 1) |
| $h_r$ | External radiative heat transfer coefficient (W/(m$^2$K)) |
| $H_{tr}$ | Heat transfer coefficient by transmission (W/K) |
| $H_{ve}$ | Heat transfer coefficient by ventilation and infiltration (W/K) |
| $I_{sol,k}$ | Solar irradiance, the total energy of the solar irradiation during the calculation period per sqm of collecting area of surface k (W/m$^2$) |
| i, k | Dummy integers |
| $N_f$ | A factor based on climate region, number of stories of a building, and sheltering from wind which is used to convert to estimated air changes in a building by natural means, without a fan (methodology from LBL) |
| $Q_{C,n}$ | Cooling need, or building energy need for cooling (MJ) |
| $Q_{C,ht}$ | Total heat transfer for the cooling mode (MJ) |
| $Q_{C,gn}$ | Total heat gains for the cooling mode (MJ) |
| $Q_{H,n}$ | Heating need, or building energy need for heating (MJ) |

| | |
|---|---|
| $Q_{H,ht}$ | Total heat transfer for the heating mode (MJ) |
| $Q_{H,gn}$ | Total heat gains for the heating mode (MJ) |
| $Q_{int}$ | Sum of internal heat gains over the given period (MJ) |
| $Q_{sol}$ | Sum of solar heat gains over the given period (MJ) |
| $Q_{tr}$ | Total heat transfer by transmission (MJ) |
| $Q_{ve}$ | Total heat transfer by ventilation (MJ) |
| $q_{ve,k}$ | Airflow rate (k element) (m$^3$/s) |
| $q_{ve,k,mn}$ | Time-average airflow rate from source k (m$^3$/s) |
| $Q_{50}$ | Air changes per hour at 50 pascal (infiltration) (ACH$_{50}$ (1/h)) |
| $R_{se}$ | External surface heat resistance of the opaque part (m$^2$k/W) |
| $SF_i$ | Total solar energy transmittance of the transparent part of the element |
| $t$ | Duration of the calculation period (Ms) |
| U | Thermal transmittance (W/m$^2$K) |
| $U_i$ | Thermal transmittance element i (W/m$^2$k) |
| $\alpha$ | Dimensionless numerical parameter depending on the time constant, $\tau$ |
| $\alpha_0$ | Dimensionless reference numerical parameter |
| $\alpha_{s,c}$ | Dimensionless absorption coefficient for solar radiation of the opaque part |
| $\gamma_C$ | Heat balance ratio for cooling |
| $\gamma_H$ | Dimensionless heat balance ratio for the heating mode |
| $\varepsilon$ | Emissivity of a surface for long-wave thermal radiation |
| $\eta_{H,gn}$ | Gain utilization factor for heating |
| $\eta_{C,ls}$ | Utilization factor for heat losses |
| $\theta_{int,set,H}$ | Set-point temperature for heating (°C) |
| $\theta_{int,set,C}$ | Set-point temperature for cooling (°C) |
| $\theta_e$ | Exterior temperature (°C) |
| $\Delta\theta_{er}$ | Average difference between the external air temperature and the apparent sky temperature (°C) |
| $\rho_a C_a$ | Heat capacity of air per volume (J/(m$^3$K)) |
| $\tau$ | Time constant of the building or building zone (h) |
| $\tau_0$ | Reference time constant (h) |
| $\Phi_{int,mn,k}$ | Heat flow gains from internal heat source k (W) |
| $\Phi_{r,k}$ | Extra heat flow due to thermal radiation to the sky from building element k (W) |
| $\Phi_{sol,k}$ | Solar heat gains through building element k (W) |
| ACH | Air Changes per Hour |
| ACH$_n$ | Natural Air Changes per Hour |
| ACH_50 | Air Changes per Hour at 50 Pascal |
| ANN | Artificial Neural Networks |
| ANOVA | Analysis of Variance |
| AR5 | Fifth Assessment Report |
| BFGS | Broyden–Fletcher–Goldfarb–Shanno Algorithm |
| CEF | Cooling Emission Factors |
| $CO_2$ | Carbon Dioxide |

| | |
|---|---|
| $CO_2eq$ | Carbon Dioxide Equivalent Emission Factor |
| COP | Coefficient of Performance |
| E | East |
| ECM | Mean Quadratic Error |
| EER | Energy Efficiency Ratio |
| EPBD | Energy Performance of Buildings Directive |
| EPW | Energy Plus Weather |
| EU | European Union |
| FA | Floor Area |
| FR | Form Ratio |
| GCM | Global Climate Model |
| GDP | Gross Domestic Product |
| GHG | Greenhouse Gases |
| GHR | Global Horizontal Solar Radiation |
| H | Hidden Layer |
| HadCM3 | Hadley Centre Coupled Model, Version 3 |
| HEF | Heating Emission Factors |
| INE | Chile Bureau of Statistics |
| IPCC | Intergovernmental Panel on Climate Change |
| IRES | International Recommendation on Energy Statistics |
| ISO | International Organization for Standardization |
| LBL | Lawrence Berkeley Laboratory |
| LCV | Low Calorific Value |
| MAE | Mean Absolute Error |
| MLR | Linear Regression Model |
| N | North |
| NE | Northeast |
| NEB | National Energy Balance |
| NS | Number of Storeys |
| NW | Northwest |
| OECD | Organization for Economic Cooperation and Development |
| Op | Opaque Surfaces |
| Op% | Percentage of Openings in the Façade |
| PM | Multilayer Perceptron Model |
| R2 | Determination Coefficient |
| RH | Relative Humidity |
| S | South |
| SA | Aysén Electrical System |
| SIC | Central Interconnected Electrical System |
| SIEC | Standard International Energy Classification |
| SING | Norte Grande Interconnected Electrical System |
| SM | Magallanes Electrical System |
| TDRe | Terms of Reference for Standardized Environmental Control and Energy Efficiency |
| UHI | Urban Heat Island |

| UNEP | United Nations Environment Programme |
| W | West |
| WMO | World Meteorological Organization |
| WWR | Window-to-Wall Ratio |

# Chapter 1
# Introduction

## 1.1 Energy in Buildings

Energy efficiency in the building sector accounts for around 30–40% of the total energy consumption of human activities as per diverse sources (Pérez-Lombard et al. 2008; UNEP 2012). In 2010, its absolute consumption was 23.7 PWh and the International Energy Agency indicates that it can reach 38.4 PWh in 2040 (IEA 2013), being responsible for 38% of the greenhouse gas emissions (UNEP 2012). Around the world, this sector currently represents 13% of the GDP and it is expected that it increases to 15% in 2020 (Global Construction Perspectives and Oxford Economics 2013). Its total budget sat at 8.2 trillion dollars in 2013 (IHS Economics 2013) and it is foreseen that this will grow to 15 trillion dollars in 2025. As such, those strategies that focused on energy efficiency, consumption and emission reduction are one of the main challenges of the construction sector. Thus, the need of predicting these factors has forced official entities, like the European Union since 2002 (European Commission 2002), to obligatorily establish the measuring of buildings' energy efficiency.

In this context, the scientific community has intensified their efforts aiming at, if not reducing, at least containing the increase of $CO_2$ emissions and energy consumption associated to the construction sector. The design, construction, operation and maintenance of a building is a long and costly process. Numerous factors are involved consuming plenty of both economic and natural resources. In the last few years, thanks to the increase in computing power, those studies focused on the simulation of building's behavior with regard to their energy demand have specially flourished. They can provide, even before starting the construction, reliable information about their energy consumption and $CO_2$ emissions during the lifespan of these buildings.

The design of a building implies plenty of parameters. The final result comes from the complex interaction of climatic variables, the building type, its shape, design, use, construction systems and air-conditioning equipment. This research covers a broad spectrum, focusing on the impact that some of these parameters could have on the energy consumption and $CO_2$ emissions. The final objective is to produce, as a result, a mathematical model or series of data that would help the designer in the early stages of the project to predict with a tolerable margin of error, the resources

© The Author(s) 2018
C. Rubio-Bellido et al., *Energy Optimization and Prediction in Office Buildings*,
SpringerBriefs in Energy, https://doi.org/10.1007/978-3-319-90146-6_1

that a building is going to deplete, thus being able to quantify its future impact on the environment.

Due to the strong influence that both local climate and local techniques exerts over the construction industry, this phenomenon is, basically, a local phenomenon. For that reason, although studies on the field of energy consumption and $CO_2$ emissions usually focus on a particular region or area, many local studies can add up to attain a more global view on this phenomenon via comparative studies. The scientific output of this research is expected to provide conclusions focused on the local case study of the Chilean context.

The amount of energy embedded in construction materials, that is to say, the energy that is consumed and the $CO_2$ that is emitted to manufacture a given material, has an impact on the total associated to the building. The results depend on the country and the context being considered. In the case of residential buildings with standardized spatial configurations that are located in Korea (Jeong et al. 2012), steel, cement and concrete cover 85% of the $CO_2$ emissions. In the case of office buildings in Greece (Dimoudi and Tompa 2008), the conclusions are similar, with the structural elements (steel, cement and concrete) being responsible for approximately 60% of the $CO_2$ emitted. In these investigations, the methodology is based on the compilation of databases that can serve as reference for designers when choosing construction materials.

Predictions of future climate scenarios and their influence on energy demand in architecture is postulated as one of the foci for research and development in the field of building science, and it has attracted increasing attention and support from governments and research institutions. Since the creation of the Intergovernmental Panel on Climate Change (IPCC) in 1988, which has recently published its Fifth Assessment Report (AR5) (IPCC 2014), there are numerous studies that consider global warming, the increase of emissions and the scarcity of natural resources. In this line, sundry prediction models have been generated for various climate scenarios (Jentsch et al. 2008). Most of these models have been developed in the United Kingdom (Mylona 2012), although they have increasingly extended through the international framework (Guan 2009; Jentsch et al. 2013). Currently, IPCC, supported by the United Nations Environment Programme (UNEP) and the World Meteorological Organization (WMO), which is the most widely accepted organization in this matter, envisages multiple emission scenarios (IPCC) for the near future (years 2020, 2050 and 2080).

Several studies have been conducted in order to assess the effect of variations on climate over the energy demand of buildings (Sorrell 2015). Due to the affinity with our case study, specific studies are cited for countries that encompass a variety of climates, because existing literature shows that a variation in the climate may have controversial effects, depending on the type of external conditions. An extensive study by Wang (Wang and Chen 2014), whose data was extrapolated to the existing building stock in the USA, proved that buildings located in temperate climates within the USA would experience an increase in energy use, basically for cooling. While, those in cold zones would reduce their energy consumption based on heating systems, due to the occurrence of warmer winters. A study from Kalvelage et al. (2014), which was focused on the influence on climate change over comfort hours and energy demand in buildings, provided data for 5 locations across the USA. Conclusions showed that, within a one-year period, the number of hours when heating is needed increases, while cooling needs decrease, and a reduction in global comfort hours is observed. Another case study referring to Spain, which also comprises a variety of climates, has been made by Gangolells and Casals (2012), with outcomes related to the energy demand of the existing building stock in the most representative cities in Spain. By using the degree-day method, it was concluded that, despite the heating demand would be reduced around 30–36% for some cities, cooling demand would rise between 107–296%. Other authors have focused on the elaboration of reliable climate files to support the simulation process. Jentsch et al. (2013) concluded that morphing actual weather files in EnergyPlus/ESP-r Weather (EPW) format with previsions from IPCC using HadCM3 files for the scenario A2, gives a consistent base to calculate variations in energy demand for buildings. These authors have also made significant contributions on the technical aspects of the morphing process of weather files (Jentsch et al. 2008), giving a reliable technical base to undertake this research.

Thus, taking into account the preceding literature, countries that have to adapt their building industry to a variety of climatic contexts may represent a compelling case study for two main reasons. First, as any given country envisages its own legal framework for this matter, they must consider all possible climatic contexts, thus dividing the territory into different zones, with controversial figures for energy demand. Secondly, future scenarios for climate change have to counterbalance the effects over different contexts, mainly regarding the energy demand for cooling and heating; in this sense, as it has been quoted before, a decrease in heating demand for cold zones may compensate the increase in cooling demand in warmer climates, or the contrary, may not be able to do so. This balance has to be taken into careful consideration, in order to determine, for any given country or territory, whether energy demand for the building sector may increase or decrease in global figures due to climate change.

In addition to the latter, it has to be taken into account that, regarding energy demand, building shape plays a crucial role. In this sense, several authors have studied the relation between changes in the future climate conditions and variations in the optimal building shape. A study by Parasonis et al. (2012) determined the relationship between the proportion and size of up to 3000 m$^2$ buildings and their energy performance, concluding that an optimization on building shape can reduce the energy demand of buildings. Other authors, such as Gong et al. (2012), have focused on various climates in China, optimizing different parameters in buildings, such as walls and roof insulation thickness, window orientation, window-to-wall ratio and glazing type, summarising seven passive design zones in which annual thermal loads can be reduced due to this optimization. A similar approach by Ihm and Krarti (2012) obtained the maximum energy saving potential in villa design in four locations in Tunisia. Other studies are specifically focused on the optimization of a single parameter, such as the window-to-wall ratio (Yang et al. 2015) in order to mitigate the impacts of climate change (Dowd and Mourshed 2015).

The need to predict energy consumption and at least contain the increase of $CO_2$ emissions in the construction sector has compelled official organizations, like the

European Union since 2002 (European Commission 2002), to order that the energy efficiency of buildings had to be quantified. One of the calculation methods recommended by the Commission Delegated Regulation (EU) No. 244/2012 (European Commission 2010) is the one contemplated in the ISO 13790:2008 (ISO 2008). This method has been widely used in the scientific community (Zhao and Magoulès 2012) in the first stages of design for both simple or complex envelopes (Negendahl and Nielsen 2015); it has even been optimized for specific climates through the use of the factor method (Jokisalo and Kurnitski 2007), as it is a validated tool and is relatively easy to make iterations, unlike the dynamic simulation methods (Negendahl 2015). When it comes to simplifying the calculation methods, there have been several investigations that relate the heating and cooling demand in regards to their energy consumption (Korolija et al. 2013a, b) and $CO_2$ emissions (Pulido-Arcas et al. 2016) with regression models.

In addition, the use of artificial neural networks has spilled over to the construction sector (Kumar et al. 2013) because they give higher feasibility and reliability than other traditional regression techniques. Most of the recent research is focused on energy used in buildings (Karatasou et al. 2006), with various applications ranging from predicting the consumption of the building per se, studies that predict energy demand (Kialashaki and Reisel 2013), energy consumption (Neto and Fiorelli 2008), both cooling and heating (Macas et al. 2016; Deb et al. 2016), and the performance of different air-conditioning systems (Kljajić et al. 2012; Rodger 2014).

Moreover, the ANN approach has allowed researchers to combine several parameters in the decision-making analyses (Cui et al. 2016) and the use of online forecasting tools (Yang et al. 2005; Li et al. 2016). These approaches are also applied to electricity load forecasting (Jurado et al. 2015), hourly energy consumption (González and Zamarreño 2005) and bioclimatic buildings (Mena et al. 2014). Other studies consider the analysis of climate parameters in the built environment as well as indoor air temperature (Ruano et al. 2006) and relative humidity (Mba et al. 2016), and there are also investigations related to thermal (Boithias et al. 2012) and visual comfort (Wong et al. 2010) levels gathered with energy consumption.

A revision of the relevant literature points out that those studies related to ANN mainly rely on the study of a single case study or a group of case studies. Mba et al. tested the performance of ANN in predicting comfort parameters in a small room ($6.5 \text{ m}^2$) located in a building in Cameroon (Mba et al. 2016). Kumar et al. proposes the application of ANN to study heating and cooling demands for a group of 250 buildings, with areas ranging from very small spaces ($1–2 \text{ m}^2$) to medium sized ones ($100 \text{ m}^2$) (Kumar et al. 2013). Karatasou et al. compared the feasibility of an ANN model against two case studies, one from a prediction tool and another from a real case study of an office building in Athens (Karatasou et al. 2006). A single case-study of an administrative building located in Sao Paulo (Neto and Fiorelli 2008) set results from a simulation software and ANN off against each over. Another single case-study was used as a test model for ANN in Southern Spain, in this case for a bioclimatic building with a peculiar energy demand (Mena et al. 2014); while, a secondary school located in Portugal was used as a test model because of its suitability for predictive neural networks (Ruano et al. 2006). A similar case-study was used as a test model for ANN for a medium sized tertiary building in Italy (Macas et al. 2016). The energy demand of three institutional buildings were assessed over 2 years and then compared with the results of an ANN model (Deb et al. 2016). A similar study proved the suitability of using Machine learning methodologies to accurately predict the energy demand of three educational buildings in Spain (Jurado et al. 2015)

With regard to tertiary buildings, Li et al. provided evidence of the adaptability of these regression techniques to predict energy demands, testing the model against two commercial premises, one small and one medium-sized property (Li et al. 2016). Energy demand for an entire rooftop air conditioning system of a single large-scale commercial building could be also accurately predicted using ANN (Rodger 2014), and similar results were provided by Kljajić et al. (2012) for a larger sample size: 65 boilers located randomly across 50 buildings in Serbia. Another case study, with regard to sample size, tested the feasibility of meta-learning based systems against 48 test buildings and 1 real building (Cui et al. 2016). Studies have been made not only for energy demand, but also for daylighting prediction using ANN, such as Wong et al. (2010), where a $35 \times 35$ m simulation of a 40 storey building was carried out as a single case study using Energy Plus software to then compare the results against ones given by the ANN model.

Other authors focus on predicting the energy demand on a large scale (nationwide study for the USA) (Kialashaki and Reisel 2013) focusing on the evolution of socioeconomic parameters or using existing data obtained from past records to predict the short-term energy consumption (González and Zamarreño 2005). Additionally, other earlier studies provide the necessary basis on how to adapt the architecture of an ANN to the requirements of a study of this kind (Yang et al. 2005) and how to improve the algorithms to better predict comfort parameters and energy consumption (Boithias et al. 2012).

Other research projects are more similar to the approach of this one, such as Khayatian et al. (2016), who uses data generated from a simulation software for energy certification in Italy as the training set in order to model an ANN that is capable of predicting the expected outcomes. Dall'O' et al. use a broad database of around 175,000 elements to provide inputs for predicting outcomes for the energy certification of buildings in a designated area (Dall'O' et al. 2015). Alternatively an ample database can be used to establish benchmarking methods for the energy performance assessment of buildings (Wang et al. 2012).

Those existing studies provide a reliable scientific corpus that widely demonstrates that ANN methodology is applicable for predicting several variables in relation to building's energy demand and energy consumption. In this way, the authors have made some contributions on this matter in the development of prediction models related to energy demand in buildings (Pulido-Arcas et al. 2016) and found out that there is room for further development in this field.

This research intends to propose an advance in the use of ANN and linear regressions to predict energy demand, energy consumption and $CO_2$ emissions, taking tertiary buildings located in Chile as a target, considering that this approach is compelling for the following reasons. At first, Chile is a country that is experimenting a profound transformation in the construction industry, as a consequence of being the first South American country to join the OECD. The legislative framework for energy efficiency has strengthened remarkably over the last few years, changing from an old 2007 standard (NCh835 2007) to some legislative texts that, despite not yet being mandatory, provide useful guidelines to reduce $CO_2$ emissions and contain energy demand both for public (MOP 2011) and residential buildings (Estándares de Construcción con Criterios de Sustentabilidad 2016). Additionally, the Chilean government has promoted a multilateral agreement with the objective of implementing the BIM (Building Information Modelling) standard in all public projects in order to improve efficiency and productivity (CORFO Chile 2016). This transformation is being done in cooperation with countries such as Spain or the UK, who provide technology and know-how about how to implement energy efficiency policies, in particular with regard to the European ISO calculation standard and energy-rating systems. Given this context, the context of this research is useful in helping this transformation into an energy-efficient and reduced-$CO_2$ policy.

## 1.2   Office Sector in Chile

Chile is a representative case study for the following reasons. This country features a particular geography, ranging from its Northernmost point 17° 29′ South to its Southernmost point 56° 32′ South, extending across 4270 km. Due to this particular geography, it encompasses a wide variety of climates, ranging, according to the Köppen-Geiger classification, from arid climates (type B) to temperate mesothermal climates (type C) and polar and alpine climates (type E). Focusing on the building sector, the current Chilean standard NCh1079:2008 (INN 2008) classifies the country into nine climatic zones, covering the aforementioned variations of Köppen-Geiger. The Chilean Ministry of Environment, in is concerns about global warming, has widely studied natural phenomena, creating the National Plan for Adaptation to Climate Change (Ministerio del Medio Ambiente Gobierno de Chile 2011) and several have come from the implications of the scientific community for these so-called futures (Eriksen et al. 2011). With regard to the building sector in Chile, energy efficiency is a relatively novel approach in public policies. The current regulation "Terms of Reference—Standardized Environmental Control and Energy Efficiency" (TDRe) (MOP 2011), which is focused on tertiary buildings, considers limitations for energy demand as well as for the comfort of their inhabitants. Within this framework, an understanding of changing climate scenarios shall provide tendencies for the energy demand of buildings, as well as the strategies upon which future revisions of the building code should be based. Thus, in a broader sense, these studies will help shape the way in which the building industry will have to reduce its dependency on energy consumption and depletion of natural resources (Robert and Kummert 2012).

In terms of the scale of the study, it has focused on tertiary office buildings, because of this sector's impact on construction activities in Chile (ERCROS 2014), comprising facilities of all kinds and types in relation to the built area, location, shape or conditioning systems, amongst other variables. According to data from the Chilean

Bureau of Statistics, office buildings are a growing sector, totalling 8.9 million square meters, approximately 12.75% of the total built area since the 2012–2015 period (INE 2015). Considering just the public policies and the Chilean Integrated Repository of Projects (BIP 2015), more than 50% of the projects currently developed by the Chilean Government are offices, and their average area is approximately 1500 $m^2$, with most of them located in the country's capital, Santiago. That is why, in order to assist this transition, this study attempts to make a relevant contribution in forecasting the three variables that are considered crucial for energy policies in the building industry: Energy demand, energy consumption and $CO_2$ emissions. As pointed out before, there is not yet a standardized quasi-dynamic calculation procedure in Chile, as in European countries, that guides the aforementioned calculations, and for this reason it has been considered a novel approach to implement the ISO 13790:2008 standard in the Chilean context.

## 1.3 Legal Framework and Energy Services

Building's $CO_2$ emissions do not just depend on the building's intrinsic parameters, as its emissions are also associated to the type of energy that this uses to cover its different demands. Thus, following the Standard International Energy Classification (SIEC) in the International Recommendation on Energy Statistics (IRES) (Overgaard 2008), we distinguish between primary and secondary energy; primary energy includes those materials that are directly burned in the building thus producing $CO_2$, such as petrol, coal and natural gas. Secondary energy includes those sources of energy, in this case electric power, which are generated elsewhere and consumed in the building. A building can consume both types of energy, so with the idea of unifying all of these into a common unit, each type of energy has a $CO_2$eq emission factor associated to quantify the emissions that their combustion or production generate. In the case of primary energy, this information is obtained directly from the type of fuel that is used (Table 1.1).

If the equipment uses electricity it is necessary to quantify the primary source that generates this, which has $CO_2$ emission factors associated, depending on its origin (wind, solar, gas, coal, nuclear, etc.). In the case of primary energy, we can take data from the international scale, as the chemical-physical characteristics of the fuels do not visibly vary from country to country; therefore, we estimate average values for the member countries of the OECD, which Chile is part of (Energy Agency) (Table 1.2).

In the case of secondary energy, it is necessary to make a distinction for each country. In the case of Chile, there are four interconnected electrical systems, the Norte Grande Interconnected System (SING by its Spanish acronym) (28.06% of the installed capacity in the country), the Central Interconnected System (SIC by its Spanish acronym) (71.03% of the installed capacity), the Aysén System (SA by its Spanish acronym) (0.29% of the installed capacity) and the Magallanes System (SM by its Spanish acronym) (0.62% of the installed capacity). Their parks are comprised

**Table 1.1** $CO_2$ emission factors associated to the different technologies used in the energy industry

| Technology | Setup | $CO_2$eq (kg/Ton) |
| --- | --- | --- |
| Petrol boilers | | 3134 |
| Diesel boilers | | 3192 |
| Liquid gas boilers | | 3042 |
| Bituminous/sub-bituminous boilers—mechanical loading from above (solid fuels) | | 2.446/1.820 |
| Boiler with mechanical loading from below (solid fuels) | | 2.455 |
| Boilers with bituminous/sub-bituminous powdered feed (solid fuels) | Dry base, lit on the sides | 2444/1819 |
| | Dry base, superficially lit | 2451/1824 |
| | Wet base | 2452/1824 |
| Boilers with mechanical loader and fluidized bed combustion chamber (Solid fuels) | Circulation bed | 2910 |
| | Effervescent bed | 2910 |
| Natural gas boilers | | 1.985 kg/m$^3$ |
| Natural gas turbines >3 MW | | 1.987 kg/m$^3$ |

*Source* Own preparation based on IPCC 2006, chart 2.7 and NEB 2009 (Emission factors and LCV obtained from IPCC 2006, density obtained from NEB 2009)

**Table 1.2** $CO_2$eq emissions factors for primary energy sources in 2015

| Source | Year (grCO$_2$eq/kWh) |
| --- | --- |
| | 2015 |
| Bituminous coal | 875 |
| Natural gas | 400 |
| Diesel | 725 |

by different types of generation, what is called the energy mix; in the SING and SM, the energy is generated by thermal plants; in the SIC, 47.41% of the generation park is from hydroelectric plants (dams and run-of-river), 51.86% by coal, petrol, diesel and natural gas combined cycle plants and 0.73% from wind farms; in the SA, 56.5% is from thermoelectric plants, 39.7% from hydroelectric and 3.8% wind. Therefore, the $CO_2$eq emissions associated to the electrical generation from each one of the interconnected systems are associated to their production characteristics (Table 1.3).

**Table 1.3** $CO_2$eq emissions for the two most important electricity production interconnected systems

| Interconnected system | Year (grCO$_2$eq/kWh) | | | | | |
| --- | --- | --- | --- | --- | --- | --- |
| | 2010 | 2011 | 2012 | 2013 | 2014 | 2015 |
| SING | 715 | 725 | 806 | 811 | 790 | 764 |
| SIC | 346 | 379 | 391 | 432 | 360 | 346 |

# References

BIP (2015) Banco Integrado de Proyectos. Ministerio de Desarrollo Social, Chile. http://bip.mideplan.cl/bip-trabajo/index.html. Accessed 20 Feb 2016

Boithias F, El Mankibi M, Michel P (2012) Genetic algorithms based optimization of artificial neural network architecture for buildings' indoor discomfort and energy consumption prediction. Build Simul 5:95–106. https://doi.org/10.1007/s12273-012-0059-6

CORFO Chile (2016) Relevante Iniciativa para Potenciar la Productividad en Infraestructura. http://www.corfo.cl/sala-de-prensa/noticias/2016/enero-2016/relevante-iniciativa-para-potenciar-la-productividad-en-infraestructura. Accessed 21 Nov 2016

Cui C, Wu T, Hu M et al (2016) Short-term building energy model recommendation system: a meta-learning approach. Appl Energy 172:251–263. https://doi.org/10.1016/j.apenergy.2016.03.112

Dall'O' G, Sarto L, Sanna N et al (2015) On the use of an energy certification database to create indicators for energy planning purposes: application in northern Italy. Energy Policy 85:207–217. https://doi.org/10.1016/j.enpol.2015.06.015

Deb C, Eang LS, Yang J, Santamouris M (2016) Forecasting diurnal cooling energy load for institutional buildings using Artificial Neural Networks. Energy Build 121:284–297. https://doi.org/10.1016/j.enbuild.2015.12.050

Dimoudi A, Tompa C (2008) Energy and environmental indicators related to construction of office buildings. Resour Conserv Recycl 53:86–95. https://doi.org/10.1016/j.resconrec.2008.09.008

Dowd RM, Mourshed M (2015) Low carbon buildings: sensitivity of thermal properties of opaque envelope construction and glazing. Energy Procedia 75:1284–1289. https://doi.org/10.1016/j.egypro.2015.07.189

IHS Economics (2013) Global construction outlook: executive outlook

Energy Agency I $CO_2$ Emissions From Fuel Combustion Highlights 2015

ERCROS (2014) Informe anual 2014. Ercros 67. https://doi.org/10.1017/cbo9781107415324.004

Eriksen S, Aldunce P, Bahinipati CS et al (2011) When not every response to climate change is a good one: identifying principles for sustainable adaptation. Clim Dev 3:7–20. https://doi.org/10.3763/cdev.2010.0060

Estándares de Construcción con Criterios de Sustentabilidad (2016) Estándares de Construcción con Criterios de Sustentabilidad. Santiago

European Commission (2002) Directive 2002/91/EC of the European Parliament and of the council of 16 December 2002 on the energy performance of buildings. Off J Eur Union 65–71. https://doi.org/10.1039/ap9842100196

European Commission (2010) Directive 2010/31/EU of the European Parliament and of the Council of 19 May 2010 on the energy performance of buildings. Off J Eur Union 13–35. https://doi.org/10.3000/17252555.l_2010.153.eng

Gangolells M, Casals M (2012) Resilience to increasing temperatures: residential building stock adaptation through codes and standards. Build Res Inf 40:1–20. https://doi.org/10.1080/09613218.2012.698069

Global Construction Perspectives and Oxford Economics (2013) Global construction 2025

Gong X, Akashi Y, Sumiyoshi D (2012) Optimization of passive design measures for residential buildings in different Chinese areas. Build Environ 58:46–57. https://doi.org/10.1016/j.buildenv.2012.06.014

González PA, Zamarreño JM (2005) Prediction of hourly energy consumption in buildings based on a feedback artificial neural network. Energy Build 37:595–601. https://doi.org/10.1016/j.enbuild.2004.09.006

Guan L (2009) Preparation of future weather data to study the impact of climate change on buildings. Build Environ 44:793–800. https://doi.org/10.1016/j.buildenv.2008.05.021

IEA (2013) World energy outlook 2013

Ihm P, Krarti M (2012) Design optimization of energy efficient residential buildings in Tunisia. Build Environ 58:81–90. https://doi.org/10.1016/j.buildenv.2012.06.012

INE (2015) Instituto Nacional de Estadísticas, Chile. http://www.ine.cl/canales/chile_estadistico/estadisticas_economicas/edificacion/series_estadisticas/series_estadisticas.php. Accessed 20 Feb 2016

INN (2008) NCh 1079. Of 2008 Arquitectura y Construcción-Zonificación climático habitacional para Chile

IPCC Data Distribution Centre. www.ipcc-data.org. Accessed 15 Feb 2016

IPCC (2014) Climate change 2014: synthesis report. Contribution of working groups I, II and III to the fifth assessment report of the intergovernmental panel on climate change

ISO (2008) EN ISO 13790: 2008 Energy performance of buildings-Calculation of energy use for space heating and cooling. 3190–200

Jentsch MF, Bahaj AS, James PAB (2008) Climate change future proofing of buildings—generation and assessment of building simulation weather files. Energy Build 40:2148–2168. https://doi.org/10.1016/j.enbuild.2008.06.005

Jentsch MF, James PAB, Bourikas L, Bahaj AS (2013) Transforming existing weather data for worldwide locations to enable energy and building performance simulation under future climates. Renew Energy 55:514–524. https://doi.org/10.1016/j.renene.2012.12.049

Jeong Y-S, Lee S-E, Huh J-H (2012) Estimation of $CO_2$ emission of apartment buildings due to major construction materials in the Republic of Korea. Energy Build 49:437–442. https://doi.org/10.1016/j.enbuild.2012.02.041

Jokisalo J, Kurnitski J (2007) Performance of EN ISO 13790 utilisation factor heat demand calculation method in a cold climate. Energy Build 39:236–247. https://doi.org/10.1016/j.enbuild.2006.06.007

Jurado S, Nebot À, Mugica F, Avellana N (2015) Hybrid methodologies for electricity load forecasting: entropy-based feature selection with machine learning and soft computing techniques. Energy 86:276–291. https://doi.org/10.1016/j.energy.2015.04.039

Kalvelage K, Passe U, Rabideau S, Takle ES (2014) Changing climate: the effects on energy demand and human comfort. Energy Build 76:373–380. https://doi.org/10.1016/j.enbuild.2014.03.009

Karatasou S, Santamouris M, Geros V (2006) Modeling and predicting building's energy use with artificial neural networks: methods and results. Energy Build 38:949–958. https://doi.org/10.1016/j.enbuild.2005.11.005

Khayatian F, Sarto L, Dall'O' G (2016) Application of neural networks for evaluating energy performance certificates of residential buildings. Energy Build 125:45–54. https://doi.org/10.1016/j.enbuild.2016.04.067

Kialashaki A, Reisel JR (2013) Modeling of the energy demand of the residential sector in the United States using regression models and artificial neural networks. Appl Energy 108:271–280. https://doi.org/10.1016/j.apenergy.2013.03.034

Kljajić M, Gvozdenac D, Vukmirović S (2012) Use of neural networks for modeling and predicting boiler's operating performance. Energy 45:304–311. https://doi.org/10.1016/j.energy.2012.02.067

Korolija I, Marjanovic-Halburd L, Zhang Y, Hanby VI (2013a) UK office buildings archetypal model as methodological approach in development of regression models for predicting building energy consumption from heating and cooling demands. Energy Build 60:152–162. https://doi.org/10.1016/j.enbuild.2012.12.032

Korolija I, Zhang Y, Marjanovic-Halburd L, Hanby VI (2013b) Regression models for predicting UK office building energy consumption from heating and cooling demands. Energy Build 59:214–227

Kumar R, Aggarwal RK, Sharma JD (2013) Energy analysis of a building using artificial neural network: a review. Energy Build 65:352–358. https://doi.org/10.1016/j.enbuild.2013.06.007

Li X, Wen J, Bai EW (2016) Developing a whole building cooling energy forecasting model for on-line operation optimization using proactive system identification. Appl Energy 164:69–88. https://doi.org/10.1016/j.apenergy.2015.12.002

M de OP (MOP) (2011) TDRe: Términos de Referencia Estandarizados con Parámetros de Eficiencia Energética y Confort Ambiental, para Licitaciones de Diseño y Obra de la Dirección de Arquitetura, Según Zonas Geográficas del País y Según Tipología de Edificios. Santiago, Chile

Macas M, Moretti F, Fonti A et al (2016) The role of data sample size and dimensionality in neural network based forecasting of building heating related variables. Energy Build 111:299–310. https://doi.org/10.1016/j.enbuild.2015.11.056

Mba L, Meukam P, Kemajou A (2016) Application of artificial neural network for predicting hourly indoor air temperature and relative humidity in modern building in humid region. Energy Build 121:32–42. https://doi.org/10.1016/j.enbuild.2016.03.046

Mena R, Rodríguez F, Castilla M, Arahal MR (2014) A prediction model based on neural networks for the energy consumption of a bioclimatic building. Energy Build 82:142–155. https://doi.org/10.1016/j.enbuild.2014.06.052

Ministerio del Medio Ambiente Gobierno de Chile (2011) Cambio climático. Informe del Estado del Medio Ambiente 2011:427–463

Mylona A (2012) The use of UKCP09 to produce weather files for building simulation. Build Serv Eng Res Technol 33:51–62. https://doi.org/10.1177/0143624411428951

NCh835 (2007) Acondicionamiento térmico-Envolvente térmica de edificios-Cálculo de resistencias y transmitancias térmicas. NCh853:24

Negendahl K (2015) Automation in Construction Building performance simulation in the early design stage: an introduction to integrated dynamic models. Autom Constr 54:39–53. https://doi.org/10.1016/j.autcon.2015.03.002

Negendahl K, Nielsen TR (2015) Building energy optimization in the early design stages: a simplified method. Energy Build 105:88–99. https://doi.org/10.1016/j.enbuild.2015.06.087

Neto AH, Fiorelli FAS (2008) Comparison between detailed model simulation and artificial neural network for forecasting building energy consumption. Energy Build 40:2169–2176. https://doi.org/10.1016/j.enbuild.2008.06.013

Overgaard S (2008) Issue paper: definition of primary and secondary energy. Prepared as input to Chapter 3: Standard International Energy Classification (SIEC) in the International Recommendation on Energy Statistics (IRES). http://unstats.un.org/unsd/envaccounting/londongroup/meeting13/LG13_12a.pdf. Accessed 5 Sep 2016

Parasonis J, Keizikas A, Kalibatiene D (2012) The relationship between the shape of a building and its energy performance. Archit Eng Des Manag 8:246–256. https://doi.org/10.1080/17452007.2012.675139

Pérez-Lombard L, Ortiz J, Pout C (2008) A review on buildings energy consumption information. Energy Build 40:394–398. https://doi.org/10.1016/j.enbuild.2007.03.007

Pulido-Arcas JA, Pérez-Fargallo A, Rubio-Bellido C (2016) Multivariable regression analysis to assess energy consumption and $CO_2$ emissions in the early stages of offices design in Chile. Energy Build 133:738–753. https://doi.org/10.1016/j.enbuild.2016.10.031

Robert A, Kummert M (2012) Designing net-zero energy buildings for the future climate, not for the past. Build Environ 55:150–158. https://doi.org/10.1016/j.buildenv.2011.12.014

Rodger JA (2014) A fuzzy nearest neighbor neural network statistical model for predicting demand for natural gas and energy cost savings in public buildings. Expert Syst Appl 41:1813–1829

Ruano AE, Crispim EM, Conceição EZE, Lúcio MMJR (2006) Prediction of building's temperature using neural networks models. Energy Build 38:682–694. https://doi.org/10.1016/j.enbuild.2005.09.007

Sorrell S (2015) Reducing energy demand: a review of issues, challenges and approaches. Renew Sustain Energy Rev 47:74–82. https://doi.org/10.1016/j.rser.2015.03.002

UNEP (2012) building design and construction: forging resource efficiency and sustainable development

Wang H, Chen Q (2014) Impact of climate change heating and cooling energy use in buildings in the United States. Energy Build 82:428–436. https://doi.org/10.1016/j.enbuild.2014.07.034

Wang S, Yan C, Xiao F (2012) Quantitative energy performance assessment methods for existing buildings. Energy Build 55:873–888. https://doi.org/10.1016/j.enbuild.2012.08.037

Wong SL, Wan KKW, Lam TNT (2010) Artificial Neural Networks for energy analysis of office buildings with daylighting. Appl Energy 87:551–557. https://doi.org/10.1016/j.apenergy.2009.06.028

Yang J, Rivard H, Zmeureanu R (2005) On-line building energy prediction using adaptive artificial neural networks. Energy Build 37:1250–1259. https://doi.org/10.1016/j.enbuild.2005.02.005

Yang Q, Liu M, Shu C et al (2015) Impact analysis of window-wall ratio on heating and cooling energy consumption of residential buildings in hot summer and cold winter zone in China. 2015

Zhao H, Magoulès F (2012) A review on the prediction of building energy consumption. Renew Sustain Energy Rev 16:3586–3592

# Chapter 2
# Research Method

## 2.1 Introduction

To evaluate the energy behavior of the buildings it is necessary to know numerous data related with its geometry, internal and external loads, construction systems, air-conditioning systems and user profiles. Selecting and quantifying the parameters needed is a complex task which requires the designer's experience and knowledge, as well as an in-depth understanding of the calculation process.

The ISO 13790:2008 standard establishes the international standard in regards to the calculation procedures to get the energy demands needed to obtain the energy consumption and $CO_2$eq emissions. This calculation method is widely seen and applied by Governments and researchers. It is recommended by the Commission Delegated Regulation (EU) No 244/2012 (EC 2012/C 115/01 2012) which implements the Energy Performance of Buildings Directive (EU 2010). This method has been widely used in the scientific community (Zhao and Magoulès 2012) in the first design stages both for simple or complex envelopes (Negendahl and Nielsen 2015), it has even been optimized for specific climates using the factor use method (Jokisalo and Kurnitski 2007), as it assumes a validated and easy-to-use tool to make iterations in contrast to the dynamic simulation methods (Negendahl 2015).

This research uses this calculation procedure for energy demand, energy consumption and $CO_2$ emissions, using an extensive series of entry data, organized into several categories, which was then introduced into equations to give the output data. Starting with a location, defined by its corresponding weather data, several prototypes are produced considering the number of stories, the area of each story, the form ratio of the buildings and the window to wall ratio. Secondly, for a given prototype building, energy demand is calculated. Thirdly, a random COP and EER is assigned in order to obtain the energy consumption for heating and cooling. Fourthly, this consumption is multiplied by the emission factor to obtain the $CO_2$eq emissions associated with the corresponding consumption.

With the data of energy demand, energy consumption and $CO_2$ emissions optimization methods are carried out. First of all, the minimal energy demand (total, cooling and heating demand) are identified considering the influence of the entry data. In a second step, energy consumption and $CO_2$ emissions are analyzed using a multivariable regression model to produce the equation that best fits its distributions

© The Author(s) 2018
C. Rubio-Bellido et al., *Energy Optimization and Prediction in Office Buildings*,
SpringerBriefs in Energy, https://doi.org/10.1007/978-3-319-90146-6_2

establishing the model's degree of accuracy. Finally, the regression models are compared with Artificial Neural Networks Model in order to best predict to predict the energy consumption and $CO_2$ emissions.

It is expected that this research assists designers in the early stages of design, being able to grasp the demand of resources of their design with a few variables. The reference values from the multivariable regressions and ANN will be of help in assisting future policy makers in establishing realistic goals in order to reduce, or at least contain, the energy demand, energy consumption and $CO_2$ emissions for office buildings.

## 2.2  Calculation Procedure

The calculation procedure used in this research is an implementation of the model defined in ISO 13790:2008 (Thermal performance of buildings—Calculation of energy use) (ISO 2008), which is aimed at performing extensive calculations for a large number of cases for the evaluation of the energy demand in non-residential buildings. It is not a tool that is aimed at energy modelling of buildings, but rather a procedure that describes buildings following a number of parameters, which are organized into 5 main groups: Location, geometry of the building, constructive systems, internal heat loads and external heat loads. Each parameter is defined by a numerical value that must be added to the calculation procedure, whose formula is also defined in the aforementioned method. Using a static simulation model and following the calculations that will be described further on, results regarding the cooling and heating demand are obtained for the studied building.

This procedure relies on a fully prescribed monthly quasi-steady calculation method, also called seasonal or monthly method (ISO 2008), that may not achieve as accurate results as the dynamic methods, present in numerous commercial simulation software. However, the ISO 13790:2008 has been duly tested with regards to its application for several European legislative frameworks with satisfactory results (Staudt and von D Hans 2010), and its accuracy is accepted for a quasi-steady calculation method, despite that each country demands some specific adaptation to their particularities. Additionally, the model considers the whole building as an interior conditioned space separated from the exterior by a thermal boundary, a single conditioned thermal zone with no internal partitions.

On the other hand, the use of this method allows for a fast and efficient modelling of a wide number of case studies, because a virtual model does not need to be assembled. By combining numerical parameters, a building prototype is correctly defined; afterwards, a wide range of case studies can be obtained by just varying those parameters within a reasonable range.

This research has implemented the calculation procedure of the ISO 13790:2008 in MS Excel® Visual basic, in order to obtain results for a great number of cases. The model runs iterations for all possible combinations of the parameters. The ISO procedure is based only on physical variables so, as the intention is to adapt this model into the Chilean context, some adaptations have been made to the original model. Thermal comfort conditions on the inside of buildings have been established following the Chilean standard, TDRe, which is specifically aimed at fostering passive building design and, therefore, reduces dependency on energy consumption. For this reason, the comfort range for occupants is extended, with the setting's operative temperature for cooling being 25 and 20 °C for heating.

### 2.2.1  Internal and External Loads

Internal heat gains in the test models are established according to the use, which is a fixed parameter for all of the prototypes; 9 h of office activity (from 8 a.m. to 5 p.m.) in which the lighting load is quantified as 16.00 W/m², occupancy loads as 6.00 W/m² and equipment loads as 4.5 W/m², according to the most representative use (Ministerio de Desarrollo Social de Chile 2016) (Fig. 2.1). The standardized and representative use for public office buildings, despite the fact this is a simplification of the real use, allow us to compare all the models with an equal pattern of internal loads.

In regards to the external heat gains, these depend on the orientation and shape of the building, thus they differ for each test model. GHR levels have been obtained from the EPW files per each location and scenario. An air flow of 0.3 l/s m² has been considered as the infiltration rate for each climate zone (Table 2.1) (Citec UBB and Decon UC 2014) considering the envelope surface and air changes per hour (Odriozoloa Maritorena 2008). The infiltration rate depends on the building compactness, considering 50 Pa air tightness of the envelope. Due to this fact, average air changes per hour (ACH) are related to air infiltration with a pressure difference of 50, which can be considered as depicted in Eq. (2.1) for large non-residential buildings.

$$ACH_{50} = \frac{Q_{50}}{60 \cdot A_E} \tag{2.1}$$

$$q_{ve,inf} = \frac{Q_{50} \cdot V}{60 \cdot A_E \cdot 3.6} \cdot N_f \tag{2.2}$$

As, according to FR, the building envelope is variable ($A_E$), the model's infiltration rate is not constant. The simulation procedure automatically transforms infiltration air limits ($Q_{50}$) following TDRe (Table 2.1) to air flow ($q_{ve,k}$ infiltration) by means of Eq. (2.2). This transformation allows us to input $Q_{50}$ to the ISO-13790:2008 procedure, as well as consider volume, $A_E$ and building exposure, which depends on the climate zone ($N_f$). Explicitly buildings, in all cases, are considered on an urban context with a constant building height of three storeys (Krigger and Dorsi 2004).

**Table 2.1** Thermal transmittance U (W/m$^2$k) and infiltrations limits (50 Pa air changes (1/h)) per climate zone

| Envelope | Climatic zones | | | | | | | | |
|---|---|---|---|---|---|---|---|---|---|
| | 1NL | 2ND | 3NVT | 4CL | 5CI | 6SL | 7SI | 8SE | 9AN |
| Roof | 0.80 | 0.40 | 0.60 | 0.60 | 0.40 | 0.40 | 0.30 | 0.25 | 0.25 |
| Wall | 2.00 | 0.50 | 0.80 | 0.80 | 0.60 | 0.60 | 0.50 | 0.40 | 0.30 |
| Storey-floor | 2.00 | 0.50 | 0.80 | 0.80 | 0.60 | 0.60 | 0.50 | 0.40 | 0.35 |
| Wall-floor | 2.00 | 0.50 | 0.80 | 0.80 | 0.60 | 0.60 | 0.50 | 0.40 | 0.30 |
| Infiltrations | 6.00 | 6.00 | 3.50 | 3.50 | 3.00 | 3.00 | 2.50 | 2.50 | 2.00 |

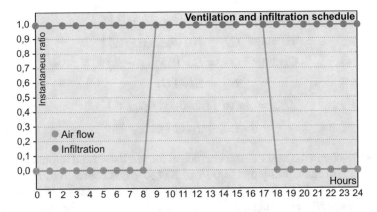

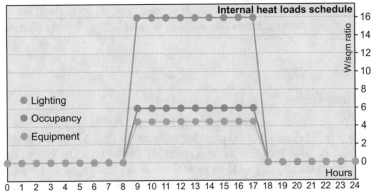

**Fig. 2.1**  Ventilation, infiltration and internal heat loads schedule of the widespread office buildings in Chile

### 2.2.2  Heat Balance

For each month, total heat balance values are obtained based on heat transfer by transmission and ventilation (Eq. 2.3). Setting temperatures used in Eqs. (2.4–2.7) are as per the model, as well as ventilation and infiltration rates (Table 2.1). The setpoint temperature for cooling 25 and 20 °C for heating $(\theta_{int,set,C}, \theta_{int,set,H})$, along with monthly outdoor average air temperatures have been extracted from Meteornom®. The values of thermal transmittance and solar factors used on transmission heat transfer calculation (Eq. 2.8) have been set at their maximum value, according to TDRe (Table 2.1), considering all the features of the building envelope. Thermal transmittance is increased 0.1 W/m² k due to thermal bridges, according to the indications from ISO-13790:2008,

$$Q_{H,ht}\ or\ Q_{C,ht} = Q_{tr} + Q_{ve} \tag{2.3}$$

$$\text{Heating:}\quad Q_{tr} = H_{tr}\left(\theta_{int,set,H} - \theta_e\right)t \tag{2.4}$$

$$\text{Cooling:} \quad Q_{tr} = H_{tr} \left(\theta_{\text{int,set,C}} - \theta_e\right) t \tag{2.5}$$

$$\text{Heating:} \quad Q_{ve} = H_{ve} \left(\theta_{\text{int,set,H}} - \theta_e\right) t \tag{2.6}$$

$$\text{Cooling:} \quad Q_{ve} = H_{ve} \left(\theta_{\text{int,set,C}} - \theta_e\right) t \tag{2.7}$$

$$H_{tr} = \Sigma i \ Ai \ Ui \tag{2.8}$$

$$H_{ve} = \rho a \ ca \left(\Sigma k \ qb_{ve,k}, kq_{ve,k,mn}\right) \tag{2.9}$$

$$q_{ve,k,mn} = f_{ve,t,k} \cdot q_{ve,k} \tag{2.10}$$

In order to obtain the heat transfer coefficient for ventilation and infiltration ($H_{ve}$) (Eq. 2.9), an air heat capacity of 1.200 J/(m³ K) ($\rho a \ ca$) is considered. The airflow $b_{ve,k}$ factor equal to 1 is estimated since the supply temperature is considered equal to outdoor temperature. Averaged airflow $q_{ve,k,mn}$, is calculated using Eq. (2.10), the time ($f_{ve,t,k}$) considered for the infiltration air rate is equal to 1, since the air flow works 24 h a day; for ventilation this value is set at 0.375, as 9 h are considered in this case.

### 2.2.3   Heat Gains

Total gains (Eq. 2.11) are obtained from internal sources (Eq. 2.12) and external solar gains (Eq. 2.13). Solar heat flow is depicted in Eq. (2.14), considering that there are no remote obstacles ($F_{sh,ob,k} = 1$) or mobile shade devices ($F_{sh,gl} = 1$), and taking into account the solar radiation that impinges on the element. It should be noted that Eqs. (2.15 and 2.16) are used for translucent and opaque elements, respectively. The absorption coefficient of the opaque surfaces ($\alpha_{s,c}$) is considered as 0.75 and the surface thermal resistance ($R_{se}$) 0.05 m² k/W. The average difference between the temperature of the outside air and the apparent temperature of the sky has been set at 11 K, in line with the standard. The exterior coefficient of heat radiation transfer ($h_r$) is estimated as 5εW/(m² K), considering emissivity of all the external enclosures as 0.9. $F_{r,k}$ equals 1 for a non-shaded flat roof and 0.5 for a non-shaded façade (Eq. 2.17).

$$Q_{C,gn} \ or \ Q_{H,gn} = Q_{int} + Q_{sol} \tag{2.11}$$

$$Q_{int} = \left(\Sigma k \ \Phi_{int,mn,k}\right) \cdot t \tag{2.12}$$

$$Q_{sol} = \left(\Sigma k \ \Phi_{sun,mn,k}\right) \cdot t \tag{2.13}$$

$$\Phi_{sol,k} = F_{sh,ob,k} \cdot A_{sol,k} \cdot I_{sol,k} - F_{r,k} \cdot \Phi_{r,k} \tag{2.14}$$

$$A_{sol,k} = F_{sh,gl} \cdot SF_i \cdot (1 - F_F) \cdot A_w \tag{2.15}$$

$$A_{sol,op} = \alpha_{s,c} \cdot R_{se} \cdot U_i \cdot A_c \tag{2.16}$$

$$\Phi_{r,k} = R_{se} \cdot U_i \cdot A_c \cdot h_r \cdot \Delta\theta_{er} \tag{2.17}$$

### 2.2.4 Energy Demand

The annual demand is obtained by adding the monthly heating and cooling requirements throughout the year; for each month. This figure is obtained from Eqs. (2.18 and 2.19), respectively.

$$\text{Heating} \quad Q_{H,n} = Q_{H,ht} - \eta_{H,gn} Q_{H,gn} \tag{2.18}$$

$$\text{Cooling} \quad Q_{C,n} = Q_{C,gn} - \eta_{C,ls} Q_{C,ht} \tag{2.19}$$

Heating demand equations consider that energy balance is produced from building heat transfer $(Q_{H,ht})$ and it is possible to reduce the need from the internal gains $(Q_{H,gn})$. However, the need for energy is directly related with the gains in cooling demand equations, and it is possible to reduce it considering building heat transfer. Hence, a utilisation factor for heating or cooling is used (Dener and Torino 2007; Jokisalo and Kurnitski 2007), $\eta_{H,gn}$, $\eta_{C,ls}$, which is in terms of the ratio of the calorific balance, $\gamma_H$, $\gamma_C$ (Eqs. 2.20 and 2.26), and a numerical parameter $\alpha$, which depend on the thermal inertia of the building (Eqs. 2.21 and 2.22), following Eqs. (2.23–2.29. The internal heat capacity of 400 kJ/m$^2$ K has been considered for the building due to its thermal inertia.

$$\gamma_H = \frac{Q_{H,gn}}{Q_{H,ht}} \tag{2.20}$$

$$\alpha = \alpha_0 + \frac{\tau}{\tau_0} \tag{2.21}$$

$$\tau = \frac{C_m/3600}{H_{tr} + H_{ve}} \tag{2.22}$$

The numerical parameter $\alpha$ considers the reference time constant $(\tau)$. For the monthly method $\tau_0 = 15$ and $\alpha_0 = 1$ are considered (Corrado and Fabrizio 2007). The utilization factor for heating or cooling can be calculated depending on the value of $\gamma_H$ or $\gamma_C$.

$$\text{If } \gamma_H > 0 \text{ and } \gamma_H \neq 1 \quad \eta_{H,gn} = \frac{1 - \gamma_H^{\alpha}}{1 - \gamma_H^{\alpha+1}} \tag{2.23}$$

$$\text{If } \gamma_H = 1 \quad \eta_{H,gn} = \frac{\alpha}{\alpha + 1} \tag{2.24}$$

$$\text{If } \gamma_H < 0 \quad \eta_{H,gn} = \frac{1}{\gamma_H} \tag{2.25}$$

The same procedure is applied for calculating the cooling demand (Eqs. 2.26–2.29).

$$\gamma_C = \frac{Q_{C,ls}}{Q_{C,ht}} \tag{2.26}$$

$$\text{If } \gamma_C > 0 \text{ and } \gamma_C \neq 1 \quad \eta_{C,ls} = \frac{1 - \gamma_C^{-\alpha}}{1 - \gamma_C^{-(\alpha+1)}} \tag{2.27}$$

$$\text{If } \gamma_C = 1 \quad \eta_{C,ls} = \frac{\alpha_C}{\alpha_C + 1} \tag{2.28}$$

$$\text{If } \gamma_C < 0 \quad \eta_{C,ls} = 1 \tag{2.29}$$

### 2.2.5 Energy Consumption and CO$_2$ Emissions

Chile features a wide variety of climates, and this is reflected in different types of environmental conditioning systems. There are locations with arid climates where only cooling equipment is needed to maintain the comfort temperatures while in cold areas, heating equipment would also be needed. In an intermediate point, there are locations where both pieces of equipment would be needed in different proportions (Rubio-Bellido et al. 2015). The different air-conditioning systems depend, in principle, on the climatic conditions, but also on the local construction industry, pursuant which there are always pieces of equipment whose use is common. Each one of the pieces of equipment, whether heating or cooling, are characterized by the COP (Coefficient Of Performance), the EER (Energy Efficiency Ratio), the HEF (Heating Emission Factor) and the CEF (Cooling Emission Factor). The first two are dimensionless factors which inform the equipment's energy efficiency and that, once introduced in the calculation model, provide the energy consumption of the equipment and, therefore, the building. The next two establish a conversion factor with the objective of transforming the energy consumed by the equipment into CO$_2$ equivalent emissions resulting from the building's air-conditioning.

A convention has been established for using this equipment. In locations where the heating demands assume at least 10% of the annual total, the building does not have specific heating equipment, and said demand is covered by the operation of cooling equipment under heat pump mode. The reason behind this decision is due to the design logic in climates with scarce heating demands, where it is more logical to acclimatize using cooling equipment through the heat pump sporadically in heating mode instead of investing in a boiler.

Once the demands are obtained, both for heating and cooling, each building is randomly assigned air-conditioning equipment pursuant the following standards. To cover the cooling demand, a piece of equipment whose EER varies randomly between 2.4 and 4.4 is assigned; and in the case of the heating, the equipment's COP varies randomly between 0.6 and 0.9. However, in case the heating demand is lower than 10% of the totals, the simulation model eliminates the choice of heating equipment and transfers the demand to the air-conditioning equipment (cooling), where an EER is assigned randomly for the heat pump mode.

The heating and cooling consumption in kWh/m$^2$ is obtained with this procedure. Finally, the $CO_2$ equivalent emissions associated to their corresponding consumptions are obtained by multiplying by the $CO_2$eq emission factor, these values are assigned randomly, as the emissions associated to the production and electricity vary every year, depending on the type of generation (energy mix).

## 2.3 Test Models

### 2.3.1 Location

Nine different locations have been considered for the test models. Each one is located in one representative city for each one of the climate zones of Chile, as defined in 1079. Of 2008. Those cities comprise a wide range of climates according to the Köppen-Geiger classification. For each location, the name of the city, the latitude and longitude and the climate zone information is provided (Table 2.3).

### 2.3.2 Geometry

Test models are considered as parallelepiped volumes with rectangular plans, located according to a perfect North-South orientation. Proportions in the three dimensions of space are defined by the gross area and number of storeys. To achieve this gross area, both the North-South and East-West façades of each test model façade vary in longitude, from 10 to 50 m each, giving as a result, multiple combinations for each direction. The parameter that relates the two of them is denoted as FR, a dimensionless coefficient. As each test model has façades in two directions, and for the sake of clarity in the discourse, FR is defined as shown in Fig. 2.2, that is, as the ratio between the length of the East-West façade and the North-South façade. For instance, an FR of 5 indicated that the East-West façade has a longitude of 50 m, and the North-South façade of 10 m. In this way, just by altering this parameter, the dimensions and orientation of the façades are precisely defined. With all the aforementioned variables, the area of the thermal enclosure (external walls, slabs and roofs) is defined.

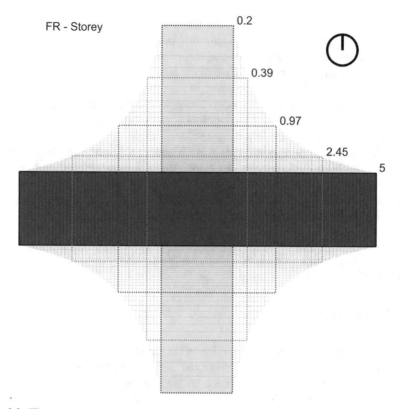

FR - Storey

0.2

0.39

0.97

2.45

5

**Fig. 2.2** FR storey

Finally, the window-to-wall ratio (WWR) coefficient is set as free, expressed as a percentage of the window area with regards to the whole area of the thermal enclosure, varying from a minimum of 10% to a maximum of 60%. These boundary limits are established following the Chilean standard, TDRe (Table 2.2), giving as a result, multiple types of façades with different WWR. Summing up, FR defines the geometrical characteristics of each test model, and WWR, their type of thermal enclosure. These two variables are set as free in the simulation and they will be the core of the analysis in this research.

## 2.3.3 Constructive Systems

The thermal envelope for each test model is established following the TDRe limit conditions (Table 2.1), which modify openings' thermal transmittance ($U_i$) and solar factor ($SF_i$) according to the window-to-wall ratio (WWR) (Table 2.2), generating 432 different cases of $U_i$ and $SF_i$. Given any location (climate zone), orientation and

**Table 2.2** Summary of base-case values for the building construction systems

| Orient. | Op. (%) | Limitation of U-value window (w/m² KJ/shading coefficient | | | | | | | | | |
|---|---|---|---|---|---|---|---|---|---|---|---|
| | | 1NL | | 2ND | | 3NVT | | 4CL | | 5CI | |
| N | 0–10 | 5.70 | – | 3.50 | – | 3.50 | – | 3.50 | – | 3.50 | – |
| N | 11–20 | 5.70 | – | 3.50 | – | 3.50 | – | 3.50 | – | 3.50 | – |
| N | 21–30 | 5.70 | – | 3.50 | – | 3.50 | – | 3.50 | – | 3.50 | – |
| N | 31–40 | 5.60 | – | 3.40 | – | 3.40 | – | 3.40 | 0.58 | 3.40 | 0.60 |
| N | 41–50 | 5.50 | 0.60 | 3.20 | 0.55 | 3.20 | 0.60 | 3.20 | 0.49 | 3.20 | 0.55 |
| N | 51–60 | 5.20 | 0.54 | 3.00 | 0.50 | 3.00 | 0.54 | 3.00 | 0.43 | 3.00 | 0.50 |
| E-W | 0–10 | 5.70 | – | 3.50 | – | 3.50 | – | 3.50 | – | 3.50 | – |
| E-W | 11–20 | 5.70 | – | 3.50 | – | 3.50 | – | 3.50 | – | 3.50 | – |
| E-W | 21–30 | 5.20 | – | 3.50 | 0.54 | 2.90 | 0.54 | 2.90 | 0.54 | 2.90 | 0.50 |
| E-W | 31–40 | 5.00 | 0.54 | 3.40 | 0.50 | 2.60 | 0.50 | 2.60 | 0.42 | 2.60 | 0.50 |
| E-W | 41–50 | 4.50 | 0.45 | 3.20 | 0.45 | 2.50 | 0.45 | 2.50 | 0.35 | 2.50 | 0.45 |
| E-W | 51–60 | 4.30 | 0.40 | 3.00 | 0.40 | 2.30 | 0.40 | 2.30 | 0.30 | 2.30 | 0.40 |
| S | 0–10 | 5.70 | – | 3.50 | – | 3.50 | – | 3.50 | – | 3.50 | – |
| S | 11–20 | 5.30 | – | 3.00 | – | 3.00 | – | 3.00 | – | 3.00 | – |
| S | 21–30 | 5.00 | – | 2.50 | – | 2.50 | – | 2.50 | – | 2.50 | – |
| S | 31–40 | 4.80 | – | 2.20 | – | 2.20 | – | 2.20 | – | 2.20 | – |
| S | 41–50 | 4.70 | – | 2.10 | – | 2.10 | – | 2.10 | – | 2.10 | – |
| S | 51–60 | 4.50 | – | 1.90 | – | 1.90 | – | 1.90 | – | 1.90 | – |
| NW/NE | 0–10 | 5.70 | – | 3.50 | – | 3.50 | – | 3.50 | – | 3.50 | – |
| NW/NE | 11–20 | 5.70 | – | 3.50 | – | 3.50 | – | 3.50 | – | 3.50 | – |
| NW/NE | 21–30 | 5.70 | – | 3.50 | – | 3.50 | – | 3.50 | 0.57 | 3.50 | – |
| NW/NE | 31–40 | 5.60 | 0.56 | 3.40 | 0.56 | 3.40 | 0.56 | 3.40 | 0.45 | 3.40 | 0.56 |
| NW/NE | 41–50 | 5.50 | 0.49 | 3.20 | 0.49 | 3.20 | 0.49 | 3.20 | 0.43 | 3.20 | 0.49 |
| NW/NE | 51–60 | 5.20 | 0.43 | 3.00 | 0.43 | 3.00 | 0.43 | 3.00 | 0.40 | 3.00 | 0.43 |
| | | 5CI | | 6SL | | 7SI | | 8SE | | 9AN | |

(continued)

**Table 2.2** (continued)

| Orient. | Op. (%) | 1NL | | 2ND | | 3NVT | | 4CL | | 5CI | |
|---|---|---|---|---|---|---|---|---|---|---|---|
| | | Limitation of U-value window (w/m² K)/shading coefficient | | | | | | | | | |
| N | 0–10 | 3.50 | – | 3.50 | – | 3.50 | 3.10 | 3.50 | – | 3.10 | – |
| N | 11–20 | 3.50 | – | 3.50 | – | 3.50 | 3.10 | 3.50 | – | 3.10 | – |
| N | 21–30 | 3.50 | – | 3.50 | – | 3.50 | 3.50 | 3.50 | – | 3.50 | – |
| N | 31–40 | 3.40 | 0.60 | 3.40 | 0.58 | 3.40 | 3.40 | 3.40 | 0.60 | 3.40 | 0.58 |
| N | 41–50 | 3.20 | 0.55 | 3.20 | 0.49 | 3.20 | 3.20 | 3.20 | 0.55 | 3.20 | 0.49 |
| N | 51–60 | 3.00 | 0.50 | 3.00 | 0.43 | 3.00 | 3.00 | 3.00 | 0.50 | 3.00 | 0.43 |
| E-W | 0–10 | 3.50 | – | 3.50 | – | 3.50 | 3.50 | 3.50 | – | 3.50 | – |
| E-W | 11–20 | 3.50 | – | 3.50 | – | 3.50 | 3.50 | 3.50 | – | 3.50 | – |
| E-W | 21–30 | 2.90 | 0.50 | 2.90 | 0.54 | 2.90 | 2.90 | 2.90 | 0.50 | 2.90 | 0.54 |
| E-W | 31–40 | 2.60 | 0.50 | 2.60 | 0.42 | 2.60 | 2.60 | 2.60 | 0.50 | 2.60 | 0.42 |
| E-W | 41–50 | 2.50 | 0.45 | 2.50 | 0.35 | 2.50 | 2.50 | 2.50 | 0.45 | 2.50 | 0.35 |
| E-W | 51–60 | 2.30 | 0.40 | 2.30 | 0.30 | 2.30 | 2.30 | 2.30 | 0.40 | 2.30 | 0.30 |
| S | 0–10 | 3.50 | – | 3.50 | – | 3.50 | 3.50 | 3.50 | – | 3.50 | – |
| S | 11–20 | 3.00 | – | 3.00 | – | 3.00 | 3.00 | 3.00 | – | 3.00 | – |
| S | 21–30 | 2.50 | – | 2.50 | – | 2.50 | 2.50 | 2.50 | – | 2.50 | – |
| S | 31–40 | 2.20 | – | 2.20 | – | 2.20 | 2.20 | 2.20 | – | 2.20 | – |
| S | 41–50 | 2.10 | – | 2.10 | – | 2.10 | 2.10 | 2.10 | – | 2.10 | – |
| S | 51–60 | 1.90 | – | 1.90 | – | 1.90 | 1.90 | 1.90 | – | 1.90 | – |
| NW/NE | 0–10 | 3.50 | – | 3.50 | – | 3.50 | 3.50 | 3.50 | – | 3.50 | – |
| NW/NE | 11–20 | 3.50 | – | 3.50 | – | 3.50 | 3.50 | 3.50 | – | 3.50 | – |
| NW/NE | 21–30 | 3.50 | – | 3.50 | 0.57 | 3.50 | 3.50 | 3.50 | – | 3.50 | 0.57 |
| NW/NE | 31–40 | 3.40 | 0.56 | 3.40 | 0.45 | 3.40 | 3.40 | 3.40 | 0.56 | 3.40 | 0.45 |
| NW/NE | 41–50 | 3.20 | 0.49 | 3.20 | 0.37 | 3.20 | 3.20 | 3.20 | 0.49 | 3.20 | 0.37 |
| NW/NE | 51–60 | 3.00 | 0.43 | 3.00 | 0.32 | 3.00 | 3.00 | 3.00 | 0.43 | 3.00 | 0.32 |

WWR, for a test model, limit values both for $U_i$ and a fixed value for the $SF_i$ are automatically established, following the TDRe limits. Thermal inertia ($C_m$) is fixed at 400 kJ/m$^2$ K (heavy construction), using Table 12 of ISO-13790:2008, which is the most commonly used and matches the ever-increasing construction techniques in Chile nowadays, due to the earthquake risk.

## 2.4 Climate Context

### 2.4.1 Current Climate Zones

The climate in Chile is very diverse, covering the climatic variants B (arid and semi-arid), C (template) and E (polar and alpine) of the Koppen-Geiger classification (Kottek et al. 2006). According to the Chilean standard TDRe, the Chilean territory is divided into 9 climatic zones, which roughly correspond with the country's 9 main climates: north-coastal, north-desert, north-transversal valleys, central-coastal, central-inland, south-coastal, south-inland, extreme south and Andean. The Chilean standard establishes said classification to facilitate a suitable conditioning of the envelope; its unification criteria are based mainly on the average temperatures of the coldest and warmest months of the year. A representative city has been chosen for each one of the 9 climate zones, which also allows symbolizing most climatic variants following the Koppen-Geiger system (Table 2.3).

EPW files have been used to model the climate of the 9 considered locations. These files include meteorological measurements for a period of 30 years featuring, among others, dry bulb temperature, relative humidity, solar radiation, wet bulb temperature,

**Table 2.3** Selected locations from Chilean Climatic Zoning

| Zone | Code | Location | Koppen-Geiger Class | Latitude | Longitude | Elevation (m) |
|------|------|----------|---------------------|----------|-----------|---------------|
| 1 | NL | Antofagasta | BWk | 23.43° S | 70.43° W | 120 |
| 2 | ND | D. de Almagro | BWk | 26.22° S | 70.03° W | 923 |
| 3 | NVT | Copiapo | BWk | 27.30° S | 70.42° W | 291 |
| 4 | CL | Valparaiso | CSb | 30.03° S | 71.48° W | 41 |
| 5 | CI | Santiago | CSb | 33.38° S | 70.78° W | 474 |
| 6 | SL | Concepción | CSb | 36.77° S | 73.05° W | 16 |
| 7 | SI | Temuco | CFb | 38.75° S | 72.63° W | 120 |
| 8 | SE | Punta Arenas | ET | 53.00° S | 70.85° W | 37 |
| 9 | AN | Lonquimay | CFb | 38.43° S | 71.23° W | 925 |

dew point and wind regime. They are widely used by many different types of computer software when applied to building energy and environmental simulation.

## 2.4.2  Climate Change Simulation

The climate in Chile is varied just like its territory, ranging from B (arid), C (temperate) to E (polar), considering the Köppen-Geiger classification (Table 1). Following the Chilean standard, 9 climatic zones have been taken into account for this research (INN 2008). For each of the 9 referenced locations, weather data files are obtained by means of Meteornom® (Table 2.3). A representative city has been selected for each of the 9 climatic zones, which comprises a wide variety of climates, following the Köppen-Geiger classification. Data is exported to EnergyPlus Weather (EPW) format. These 9 files have been used to model the so-called "base scenarios", whose data files comprise average climate values for the 1960–1991 period; from these, data regarding Global Horizontal Solar Radiation (GHR, Wh/m$^2$), Relative Humidity (RH, %) and Dry Bulb Temperature (DBT, °C) has been selected.

Future climate scenarios have been modelled by means of the UK Met Office Hadley Centre Coupled Model 3 HadCM3 (Met Office Hadley Centre 2016). This model takes into account the combination of A2a, A2b and AS2c scenarios in regards to $CO_2$ emissions. Using the morphing tool, CCWorldWeatherGen (Met Office Hadley Centre 2016), based on the studies of Belcher et al. (2005), the EPW files for the 9 base scenarios are morphed with the GHG A2 emissions scenario, obtaining sets of data for the years 2020, 2050 and 2080. Thus, in total, 27 climate future scenarios have been produced and altogether, 36 climate scenarios compose the database used in this research.

The authors consider that the aforementioned predictions can be considered as a likely future development for this time-span. Although the HadCM3 A2 model is the most extensive to perform simulations for future predictions, and also, in spite of the numerous research papers that rely on its basis, some considerations should be made with regard to its accuracy and limitations.

- A certain uncertainty in the resolution of the Coupled Model (GCM) is found for specific cases. The HadCM3 model has a grid resolution of $2.5 \times 3.8°$(Pope et al. 2000), which means that each simulated grid element should cover around $278 \times 422$ km. Although this model can be applied to this case-study, it is primarily used for making predictions at a global scale, and that gives some uncertainty for objects of study whose size is relatively small compared to the grid resolution.
- EPW files are composed by measurements from meteorological stations, thus provide open field conditions. That is, they do not envisage specific contour conditions that can alter energy demand for buildings, such as solar obstructions, urban heat island (UHI) or microclimate effect.

– HadCM3 models foresee future trends for the average values regarding climate variables, but it is not capable of contemplating extraordinary natural phenomena associated with climate change: heavy seasonal floods, hurricanes, storm surges, periods of drought, etc.

That said, the future scenarios that are used in this research should be understood as the most probable expected average climatic conditions for a given time-span and for open field boundary conditions, in selected locations with a margin of error that corresponds to the grid size of the HadCM3 model.

## 2.5 Optimization and Prediction Methods

### 2.5.1 Minimal Energy Demand

Multiple iterations based on the 5 main categories (location, geometry, constructive systems, internal and external heat loads) are made according to the Chilean standard TDRe and ISO-13790:2008 for each location and scenario. For each climate zone, simulations are set considering the climate change scenarios. An algorithm is implemented in the script in order to find the combination of building parameters that give as a result the minimal energy demand per each scenario. This process is divided in two phases. First, the algorithm finds the best possible combination of both current and future energy demand, considering heating and cooling demand as a whole. Then, the algorithm analyses again the original dataset, but this time making a difference between cooling and heating demand, in order to clarify how that best possible combination that was found in the first phase balances its energy demand between heating and cooling. This approach allows to see at first how the buildings will behave in general terms, considering energy demand as a whole. After that, the optimization method goes further deep and splits the demand into heating and cooling in order to clarify which one is more influential. In this way, the optimization method is based on dataset results that allows understanding the evolution of annual global energy demand together with heating and cooling over time and their implication on building shape and form.

### 2.5.2 Multiple Linear Regressions

Given the quantitative variable Y and the set of p predictor variables $X_1 \ldots X_p$, the Multiple Linear Regression (MLR) model assumes that the mean of Y (Eq. 2.30) determines the values of the predictor variables in a linear combination:

$$Y = \beta_0 + \beta_1 X_1 + \cdots + \beta_p X_p + \varepsilon \tag{2.30}$$

MLR is a classic technique that provides several advantages: simplicity, interpretability, possibility of being adjusted over the transformations of the variables, and the performing of reasoning, supposing the hypothesis of normality, homoscedasticity and intercorrelation between the error $\varepsilon$ and the predictor variables. In this work, the function lm of the package R (R Core Team, 2016) has been used for the adjustment by minimum blocks of the MLR model. Multiple linear regression models do not require any additional configuration for the parameters by means of validation process, thus they were adjusted on the basis of the conjunction of both validation and training sets, being later applied to the test data.

### 2.5.3   Multilayer Perceptron

The Artificial Neural Network (ANN) is a computational paradigm which provides a great variety of mathematical non-linear models, which are useful for tackling different statistical problems. Several theoretical results support a particular architecture, namely the multilayer perceptron (PM), for example, the universal approximate property, as in Bishop (1995). We have considered a three-layered perceptron with the logistic activation function $g(u) = e^u/(e^u + 1)$, in the hidden layer, and the identity function as the activation function for the output layer. By denoting H as the size of the hidden layer, $\{v_{ih}, i = 0,1,2,\ldots, h = 1,2,\ldots,H\}$ as the synaptic weights for the connections between the p-sized input and the hidden layer, and $\{w_h, h = 0,1,2,\ldots,H\}$ as the synaptic weights for the connections between the hidden and the output layer, then the output of the neural network from a vector of inputs $(x_1,\ldots,x_p)$ becomes (Eq. 2.31):

$$o = w_o + \sum_{h=1}^{H} w_h g \left( v_{0h} + \sum_{i=1}^{p} v_{ih} x_i \right) \tag{2.31}$$

The net R function (Venables and Ripley 2002) fits single hidden-layer neural networks by using the Broyden–Fletcher–Goldfarb–Shanno algorithm (BFGS) procedure, a quasi-Newton method also known as a variable metric algorithm, which attempts to minimize a least-square criterion which introduces a decay term $\lambda$ in an effort to prevent problems of overfitting. The BFGS algorithm can be found in Bishop (1995). Defining $W = (W_1,\ldots,W_M)$ as the vector of all the M coefficients of the net, and given n target values $y_1,\ldots,y_n$, the M parameters are estimated through the following optimization problem (Eq. 2.32):

$$\underset{W}{\text{Min}} \sum_{i=i}^{n} \left\| z_i - \hat{z}_i \right\|^2 + \lambda \left( \sum_{i=i}^{M} W_i^2 \right) \tag{2.32}$$

It is well known that the performance of the final model can be improved with the normalization of the input variables. In this way, each variable was transformed to achieve a mean equal to zero and a standard deviation equal to 1. We must note that this transformation worked out with the means and standard deviations computed on the training set, and those statistics were used to normalize the test set.

The implementation of a PM model requires the specification of two parameters: the size of the hidden layer (H) and the decay parameter, and therefore a 10-fold cross-validation search was carried out with the tune.nnet function in R over a grid defined as $\{1,2,3,\ldots,10\} \times \{0,0.01,0.05,0.1,0.2,\ldots,1.5\}$. Amongst all possible pairs, the one with the lowest Mean Quadratic Error (ECM) value was selected in the validation set, obtaining always null (0) as the regularization parameter. The latter agreed with the fact that the big size of the training set and the reduced number of predictors does not seem to be a prone scenario for overfitting problems. Once the suitable size for the hidden layer was identified, the multilayer perceptron was trained using the function tune.nnet in R, over the junction of the training and validation sets. Finally, each compiled PM model was applied over the test data, in order to obtain reliable estimations regarding the possibility of generalization.

# References

Belcher SE, Hacker JN, Powell DS (2005) Constructing design weather data for future climates. Build Serv Eng Res Technol 1:49–61. https://doi.org/10.1191/0143624405bt112oa

Bishop CM (1995) Neural Networks for Pattern Recognition, 1st edn. Oxford University Press, New York

Citec UBB, Decon UC (2014) Manual de Hermeticidad al aire de Edificaciones. (MOP), Ministerio de Obras Públicas

Corrado V, Fabrizio E (2007) Assessment of building cooling energy need through a quasi-steady state model: Simplified correlation for gain-loss mismatch. Energy Build 39:569–579. https://doi.org/10.1016/j.enbuild.2006.09.012

Dener E, Torino P (2007) Building energy performance assessment through simplified models: application of the ISO 13790 quasi-steady state method. Build Simul 2007:79–86

EC 2012/C 115/01 (2012) Commission Delegated Regulation (EU) No 244/2012 of 16 January 2012 supplementing Directive 2010/31/EU of the European Parliament and of the Council on the energy performance of buildings by establishing a comparative methodology framework for calculating. Off J Eur Union 28. https://doi.org/10.3000/1977091x.c_2012.115.eng

Eu (2010) Directive 2010/31/EU of the European Parliament and of the Council of 19 May 2010 on the energy performance of buildings. Off J Eur Union 13–35. https://doi.org/10.3000/17252555.1_2010.153.eng

INN (2008) NCh 1079. Of 2008 Arquitectura y Construcción- Zonificación climático habitacional para Chile

ISO (2008) EN ISO 13790: 2008 Energy performance of buildings-Calculation of energy use for space heating and cooling. 3190–200

Jokisalo J, Kurnitski J (2007) Performance of EN ISO 13790 utilisation factor heat demand calculation method in a cold climate. Energy Build 39:236–247. https://doi.org/10.1016/j.enbuild.2006.06.007

Kottek M, Grieser J, Beck C et al (2006) World map of the Köppen-Geiger climate classification updated. Meteorol Zeitschrift 15:259–263. https://doi.org/10.1127/0941-2948/2006/0130

Krigger J, Dorsi C (2004) Residential energy: cost savings and comfort for existing buildings, 4th edn. Saturn Resource Management, Inc., Helena

Met Office Hadley Centre (2016) Met Office Hadley Centre for climate science and services. http://
    /www.metoffice.gov.uk/climate-guide/science/science-behind-climate-change/hadley. Accessed
    20 Feb 2016
Ministerio de Desarrollo Social de Chile (2016) Banco Integrado de Proyectos. Ministerio de Desar-
    rollo Social, Chile. http://bip.mideplan.cl/bip-trabajo/index.html. Accessed 20 Feb 2016
Negendahl K (2015) Automation in construction building performance simulation in the early
    design stage: an introduction to integrated dynamic models. Autom Constr 54:39–53. https://doi.
    org/10.1016/j.autcon.2015.03.002
Negendahl K, Nielsen TR (2015) Building energy optimization in the early design stages: a sim-
    plified method. Energy Build 105:88–99. https://doi.org/10.1016/j.enbuild.2015.06.087
Odriozoloa Maritorena M (2008) Cálculo y medida de inflitraciones de aire en edificios. Universidad
    del País Vasco
Pope VD, Gallani ML, Rowntree PR, Stratton Ra (2000) The impact of new physical parametriza-
    tions in the Hadley Centre climate model: HadAM3. Clim Dyn 16:123–146. https://doi.org/10.
    1007/s003820050009
Rubio-Bellido C, Pulido-Arcas JA, Ureta-Gragera M (2015) Aplicabilidad de estrategias genéri-
    cas de diseño pasivo en edificaciones bajo la influencia del cambio climático en Concepción y
    Santiago, Chile. Hábitat Sustentable 5:33–41
Staudt A, van D Hans ED (2010) Report on the application of CEN standard EN ISO 13790: energy
    performance of buildings—calculation of energy use for space heating and cooling
Venables WN, Ripley BD (2002) Modern applied statistics with S, 4th edn. Springer, New York
Zhao H, Magoulès F (2012) A review on the prediction of building energy consumption. Renew
    Sustain Energy Rev 16:3586–3592

# Chapter 3
# Energy Demand Analysis

## 3.1 Introduction

The energy demand analysis is performed following a methodology that comprises three main stages (Fig. 3.1). In the first stage, the input data for the simulation process is set up, making a distinction between two groups: Climate data and what has been called "test models". On one side, the 9 different climate zones in which Chile is divided into by the Chilean building standard are considered; these zones cover all the existing climate contexts in the country and have been called "climate scenarios". For each of these 9 zones, files containing the current climate data have been compiled. These files have been "morphed" according to the predicted climate scenarios for 2020, 2050 and 2080, producing a new set of climate files for these future years. These files will be used as the external conditions for the calculation of the energy demand. On the other side, test models have been defined following the parameters of the TDRe standard; some variables have been fixed while those related to the building shape and the enclosure will be set as free and studied in this research.

In the second stage, all the input data is input into the simulation routine, whose calculation model is based on the ISO-13790:2008 procedure, aimed at obtaining energy demand in buildings. These calculations have been performed for numerous variations of the selected variables from the first stage. As a result, output data regarding energy demand is obtained for a combination of parameters.

In a third stage, the output data is analysed in three main groups. First, the morphed climate files are analysed to clarify which changes are expected to occur in the near future or, in other words, what the tendency regarding climate change will be. Secondly, the expected future energy demand for different combinations of building shapes and enclosures will be clarified. In a later stage, this data will be analysed in order to find the most optimal combination of building shape and enclosure for each given climate scenario, with the objective of establishing design strategies.

© The Author(s) 2018
C. Rubio-Bellido et al., *Energy Optimization and Prediction in Office Buildings*,
SpringerBriefs in Energy, https://doi.org/10.1007/978-3-319-90146-6_3

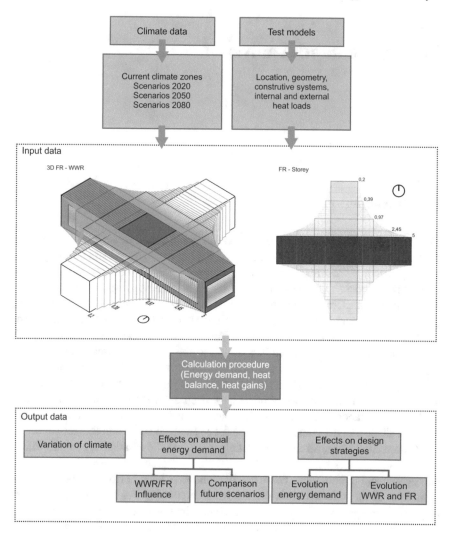

**Fig. 3.1** Methodology for annual energy demand optimization under the impact of climate change

## 3.2   Climate Variation

The original EPW files provide the average climate conditions for the 9 considered locations for the base scenario. After applying the morphing procedure under the GHG A2 scenario, these EPW files are transformed to represent the forecasted average climate conditions in 2020, 2050 and 2080. From all the climatic variables, the focus is placed on Global Horizontal Solar Radiation (GHR, Wh/m$^2$), Relative Humidity (RH, %) and Dry Bulb Temperature (DBT, °C). For each location, a comparison is set between the base scenario and the three future scenarios, leading to a clarification regarding the change of the aforementioned variables. Each of these variables comprises 8760 items along with hourly data that comprises the whole year. For greater clarity, these predictions have been grouped by year (Table 3.1) and by month (Fig. 3.2).

With regard to these changes, the following tendencies can be outlined. First, average Dry bulb temperature (DBT, °C) increases in all locations. However, the largest increases are predicted in warmer climates: 4.19 °C in 2080 in 1NL, 3.73 °C in 2ND, 3.57 °C in 3NVT, 3.18 °C in 4CL and 3.67 °C in the 5CI zone, whereas in cold climates dry bulb temperatures would increase 2.55 °C in 6SL, 2.50 °C in 7SI and 2.14 °C in 8SE zone. A singular pattern occurs in the 9AN zone, where DBT would increase by 3.29 °C. Thus, in general terms, all climates would become warmer when comparing to the present situation. Relative humidity (RH, %) would decrease in all zones, so that all climates would become drier with the following figures: −4.17% in 2080 in 1NL, −4.58% in 2ND, −3.51% in 3NVT, −5.42% in 4CL, −5.92% in 5CI zone, −2.83% in 6SL in −2.50% 7SI and −0.57% in the 5CI zone. Again, 9AN presents a singular pattern, with a decrease of 5.16%. Global Horizontal Solar Radiation (GHR, Wh/m$^2$) is not greatly altered according to AR5 (IPCC), with slight increases or decreases in each climate zone in 2080 for the A2 scenario: 0.67% (1NL), −1.87% (2ND), −3.46% (3NVT), 8.48% (4CL), 8.62% (5CI), 10.75% (6SL), 14.40% (7SI), −8.02% (8SE) and 16.10% (9AN).

## 3.3  Effects on Annual Energy Demand

### 3.3.1  WWR and FR Influence

The simulation routine has considered all the possible combinations of the variables that were set as free, following Table 3.2, that is FR, which varies from 0.2 to 5 in 41 steps and WWR, which varies from 10 to 60%, in steps of 1%. From all the possible variables combinations, 2050 cases are obtained for each year of study per each location. Covering the four scenarios, 8200 cases are considered for each location, with a total of 73,800 cases. Although the simulation model performs analysis for WWR variations in 1% steps, with the lower and upper limit values of 10–60% respectively, for the sake of clarity, the annual energy demands is depicted for the intervals 10, 20, 30, 40, 50 and 60% (Fig. 3.3).

For each climate zone and climate scenario, the optimized values for WWR and FR have been identified. This term is defined as the single value or combination of values for WWR and FR that achieves the minimum energy demand. With this in mind, when speaking about the 9 different locations, the outcomes from the simulations can be brought into four categories. The first group are those climatic zones where the optimized WWR and FR are similar for the four considered scenarios (3NVT). The second group are those areas where the optimal WWR remains constant in all scenarios, but optimal FR differs (2ND, 4CL, 5CI, 6SL 8SE and 9AN). The third

**Table 3.1** Annual average for dry bulb temperature, relative humidity and global horizontal radiation "Current", 2020, 2050, 2080 years under the GHG A2 scenario in the 9 climate zones of Chile

| Parameter | Year | Code zone | | | | | | | | |
|---|---|---|---|---|---|---|---|---|---|---|
| | | 1 N | 2ND | 3NVT | 4CL | 5CI | 6SL | 7SI | 8SE | 9AN |
| Dry bulb temperature (°C) | Current | 16.46 | 18.36 | 17.34 | 15.90 | 14.70 | 12.65 | 11.40 | 6.37 | 11.16 |
| | 2020 | 17.44 | 19.28 | 18.23 | 16.79 | 15.70 | 13.41 | 12.20 | 7.03 | 12.16 |
| | 2050 | 18.64 | 20.30 | 19.24 | 17.69 | 16.81 | 14.07 | 12.82 | 7.59 | 13.05 |
| | 2080 | 20.65 | 22.09 | 20.91 | 19.08 | 18.37 | 15.20 | 13.90 | 8.51 | 14.45 |
| Relative humidity (%) | Current | 76.00 | 65.92 | 69.11 | 64.39 | 65.39 | 79.31 | 79.82 | 73.83 | 63.72 |
| | 2020 | 73.34 | 63.58 | 67.52 | 62.72 | 63.80 | 78.64 | 78.74 | 73.75 | 62.13 |
| | 2050 | 72.09 | 62.16 | 65.98 | 60.56 | 60.98 | 77.81 | 78.32 | 73.46 | 60.46 |
| | 2080 | 71.83 | 61.34 | 65.60 | 58.97 | 59.47 | 76.48 | 77.32 | 73.26 | 58.56 |
| Global horizontal radiation (Wh/m$^2$) | Current | 439.84 | 447.30 | 437.95 | 310.23 | 388.60 | 339.33 | 290.02 | 225.09 | 346.40 |
| | 2020 | 441.55 | 448.67 | 436.60 | 311.38 | 389.62 | 342.48 | 297.13 | 223.70 | 351.00 |
| | 2050 | 441.98 | 449.35 | 438.49 | 316.07 | 396.24 | 344.62 | 297.66 | 220.34 | 355.71 |
| | 2080 | 440.51 | 445.43 | 434.49 | 318.71 | 397.22 | 350.08 | 304.42 | 217.07 | 362.50 |

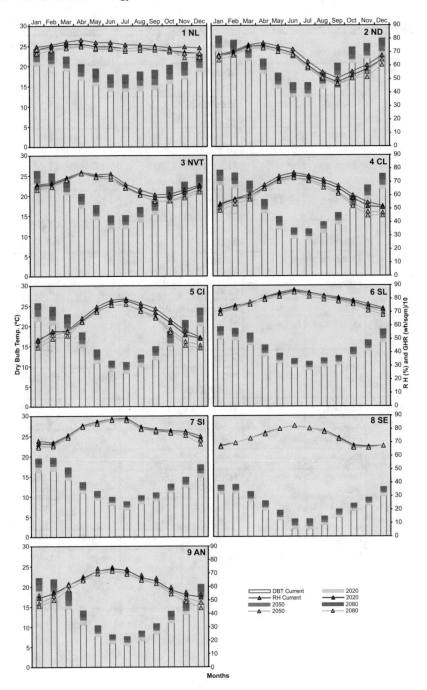

**Fig. 3.2**   Average dry bulb temperature (DBT) and relative humidity (RH) for the current scenario, 2020, 2050, 2080 under the A2 (medium-high) emissions scenario in the 9 climate zones of Chile

**Table 3.2** Parameters that define each test item for energy demand building modelling

| | | Units | Source | Lower limit | Upper limit | Number of items |
|---|---|---|---|---|---|---|
| Location | City | – | Manual input | | | 9 |
| | Lat. and long. | Degrees | Meteonorm | | | 9 |
| | Climatic zone | – | 1079. of 2008 | | | 9 |
| Geometry | Gross area | $m^2$ | Manual input | 1500 | 1500 | 1 (fixed) |
| | Storeys | integer | Manual input | 3 | 3 | 1 (fixed) |
| | North façade | m | Manual input | 10 | 50 | 40 |
| | East façade | m | Manual input | 10 | 50 | 40 |
| | FR | – | Calculated | 0.2 | 5 | 41 |
| | WWR | % | TDRe | 10 | 60 | 50 |
| Constructive systems | $U_i$ | $W/m^2 K$ | TDRe | 0.4 | 5.7 | 432 |
| | $SF_i$ | – | TDRe | 0 | 9 | 432 |
| | $C_m$ | $kJ/m^2 K$ | ISO 13790:2008 | 400 | 400 | 1 (fixed) |
| Internal heat loads | Lighting | $W/m^2$ | TDR | 16 | 16 | 1 (fixed) |
| | Occupancy | $W/m^2$ | TDRe | 6 | 6 | 1 (fixed) |
| | Equipment | $W/m^2$ | TDRe | 4.5 | 4.5 | 1 (fixed) |
| | Intensity of use | h | TDRe | 9 | 9 | 1 (fixed) |
| External heat loads | GHR | $W/m^2$ | From EPW | 19.16 | 617.79 | 2160 |
| | Air flow | $l/s*m^2$ | TDRe | 0.3 | 0.3 | 1 (fixed) |
| | Infiltration rate | $ACH_{50}$ | TDRe | 2.0 | 6.0 | 351 |
| | Conversion factor | $ACH_{50}$ to $ACH_n$ | LBL | 10.9 | 17.2 | 4 |

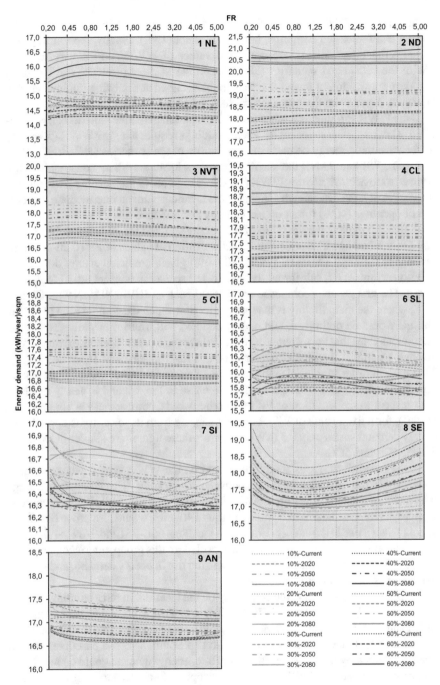

**Fig. 3.3**  FR and WWR (every 10%) versus energy demand for current scenario and 2020, 2050, 2080 under the A2 (medium-high) emissions scenario in the 9 climate zones of Chile

group consists of the areas in which recommended FR is similar in regards to the current and forecasted scenarios, but WWR is not (1NL). Finally, the fourth group includes those locations where optimized values of WWR and FR vary for each one of the considered scenarios (7SI). According to these results, it can be stated that the optimization of the geometric variables, aimed at achieving minimal energy demands, depends on the geographical location and the climate conditions.

If the WWR variation is analyzed as a single parameter, three patterns can be identified. The first refers to locations where buildings require a higher WWR (50–60%) to decrease their energy demand in the majority of the scenarios (1NL, 2ND, 4CL, 5CI, 6SL, 7SI and 9AN); the second comprises locations where constructions are compelled to achieve a lower WWR (10–20%) to improve their energy performance (8SE). Finally, the last category includes the locations with an intermediate percentage of openings (40%) for an optimal energy performance (3NVT).

Figure 3.3 depicts the variation of the energy demand for all the possible WWR and FR combinations. Given a fixed location, different values of FR (0.20–5), combined with different values of WWR (10–60%) give a predicted energy demand in (kWh/year)/m$^2$ as a result for all the considered scenarios (current, 2020, 2050 and 2080). The first group is composed of locations such as 1NL, 2ND, 3NVT, 4CL and 5CL, where the main finding is that, no matter how much the combination of WWR and FR would be optimized, and energy demand will irretrievably rise. The set of continuous lines that corresponds to the 2080 scenario presents energy demand values that are higher than their counterparts; this rise ranges about 1 (kWh/year)/m$^2$. That said, in order to minimize the increase of energy demand, some trends can be outlined. In all locations except for 1NL, the rise in the demand is nearly inelastic with respect to the variation of FR or, in other words, compactness is not an issue. In 2ND, for a fixed value of WWR and FR varying from its lower to its upper limit, energy demand varies only 0.5 (kWh/year)/m$^2$; these figures are similar in locations 3NVT, 4CL and 5CL.

In these locations, focus should be placed on the WWR values, whose optimal values are in the intermediate range, between 40–60%. In 1NL, compact buildings with an FR of around 0.45–1 present a higher energy demand in all climate scenarios. Besides, an intermediate WWR ratio, of around 40–60% would also prove beneficial to counteract a rise in the energy demand. In locations 6SL, 7SI and 9AN the situation is unclear, at first, but some trends can be clarified. In this group, intermediate WWR values (40–60%) are the key factor to contain the rise in energy demand, and the best value at these locations is clearly 60%.

FR plays a contradictory role depending on the location. In 6SL, buildings should be compact (0.7–1.2) and have a WWR of 60% to achieve a minimum energy demand. In 7SI and 9AN FR, the upper value of 5 should be reached, meaning that the building should face its longest façades to the North and South. 8SE is a particular case in itself, explained by its particular location. Here, energy demand could be reduced if a proper combination of WWR and FR is adopted. In all scenarios, it is clear that compactness (FR 0.8–1.25) allows for a lower energy demand, so the key factor in this case is WWR, whose optimal values should be the lowest possible (10%).

### 3.3.2 Annual Energy Demand for Different Climate Scenarios

For each location, the 2050 possible combinations of energy demand have been assessed for the current scenario, 2020, 2050 and 2080. Going one step further in the discussion, they are compared as a whole, trying to attain the lowest possible energy demand (Fig. 3.4). Overall, it can be stated that climate change will have significant consequences on the energy performance of existing and future buildings. The results regarding energy demand can be gathered in two categories. First, the climate zones where energy demand will increase: 1NL, 2ND, 3NVT, 4CL, 5CL and 9AN; the second category comprises the zones where energy demand is expected to decrease: 6SL, 7SI and 8SE. For each climate scenario, the optimized energy demand is outlined for the sake of clarity in the discourse.

There are locations where the difference between the optimized current demand and the optimized demand in 2080 is less than 0.75 $(kWh/year)/m^2$: (1NL, 6SL, 7SI, 8SE and 9AN); in other locations the difference is greater than 1.50 $(kWh/year)/m^2$, even reaching a difference of 3.266 $(kWh/year)/m^2$ in the location 2ND. Hence, geometry optimization has a decisive role in coastal Northern, Southern and Andean areas of Chile. Various locations are easily predictable following the building performance, since all the cases depict the same trend, implying either a reduction or an increase in energy demand (1NL, 2ND, 3NVT, 4CL, 5CI, 8SE, 9AN). This situation is harder to predict in 6SL and 7SI, because in these zones, according to the outcomes produced by the model, energy demand is, whether increased or steadily maintained, based on the considered geometrical characteristics (combination of FR and WWR).

There is a wide variability of performance depending on the climatic zone. In 3NVT, 4CL, 5CI and 8SE, it can be appreciated that although energy demand varies for different geometries, similar values are found in the 2020, 2050 and 2080 scenarios. However, in zones 2ND, 6SL, 7SI and 9 AN, 2020 and 2050 scenarios present a common trend, marking clear differences in performance compared with the 2080 scenario. Energy demand in 1NL is reduced by 2020 and 2050, however, it will tend to increase for the year 2080.

### 3.3.3 Heating and Cooling Energy Demand for Different Climate Scenarios

The energy demand combinations discussed above (Fig. 3.4) correspond to the sum of the demand for heating and cooling for the current time, 2020, 2050 and 2080. To have a better view of the variation, both demands have been mapped independently, making the combination for the heating and cooling demands of the optimized annual energy demand (Fig. 3.5). In general, it can be said that climate change will lead to an increase of cooling demands and decrease in heating demands. Specifically, for the selected cases, heating minimum demands will be reduced by between 2.62 and 0.54

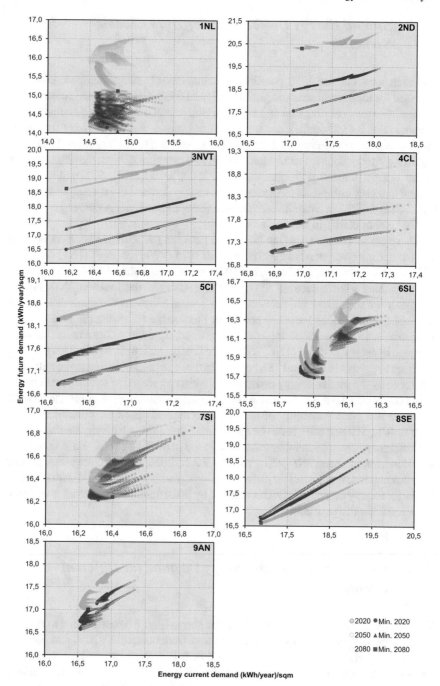

**Fig. 3.4** Current energy demand versus future energy demand for 2020, 2050, 2080 under the A2 (medium-high) emissions scenario in the 9 climate zones of Chile

(kWh/year)/m$^2$, while cooling will increase between 4.47 and 0.53 (kWh/year)/m$^2$. The results discussed in the previous section were grouped into climatic zones where the annual energy demand r will increase (1NL, 2ND, 3NVT, 4CL, 5 CL and 9AN); and into areas where it is expected that the annual energy demand decreases (6SL, 7SI and 8SE). This is because the areas in which cooling demand increases due to the increase of outdoor temperatures is much greater than the reduction that heating demands will have.

Analyzing the differences between 2080 and today, it is seen that the climatic zones which will produce a higher increment of cooling demand are 2ND and 3NVT, which will increase in the selected cases by 4.47 and 3.50 (kWh/year)/m$^2$ while the heating demand will be reduced in 1.21 and 1.02 (kWh/year)/m$^2$. This entails an increase in the total annual demand of 3.26 and 2.48 (kWh/year)/m$^2$, respectively.

4CL and 5CL zones also suffer increases in cooling demands (2.76 and 2.85 (kWh/year)/m$^2$), with these being significantly less than the previous zones. The heating demand will be reduced in 1.17 and 1.27 (kWh/year)/m$^2$, which indicates that the increase in total annual energy demand is lower. The Coastal-North and Andean regions (1NL and 9AN) also increase total demands but, with a different pattern as can be seen in Fig. 3.3. The heating energy demand of 1NL zone will be reduced by 2.62 (kWh/year)/m$^2$, while in the 9AN 0.59 (kWh/year)/m$^2$, this similarly occurs with cooling demands which increase by 3.25 and 1.07 (kWh/year)/m$^2$ respectively, representing 0.63 and 0.48 (kWh/year)/m$^2$ of the total increase.

Climatic zones in which the increase of outdoor temperatures scarcely affects optimal cases are 6SL, 7SI and 8SE. In these areas, there will be an increase in cooling demands (0.59, 0.53, 0.58 (kWh/year)/m$^2$) and the reduction of heating demands are similar (0.725, 0.55, and 0.85 (kWh/year)/m$^2$), in all cases having a greater heating rather than cooling reduction. Therefore, the reduction in the annual energy demand is minimal, ranging between 0.02 and 0.27 (kWh/year)/m$^2$.

## 3.4  Effects on Design Strategies

### 3.4.1  Evolution on Annual Energy Demand

Taking as a base all 73,800 simulated test models, optimized energy demands in each of the 9 locations, for each climate scenarios, have been extracted from Figs. 3.4 and 3.5 and depicted in Fig. 3.6. Taking into account only these optimized cases, there are locations where, even the best optimized buildings will inevitably increase their annual energy demand (1NL, 2ND, 3NVT, 4CL, 5CI, 9AN). However, there are other locations, such as Concepción (6SL), Temuco (7SI), and Punta Arenas (8SE), where the energy demand could be noticeably close to current figures according to the forecasted 2020, 2050 and 2080 scenarios, if the best combination of WWR and FR is achieved, and may be reduced between 0.13 (kWh/year)/m$^2$ (6SL) and 0.27 (kWh/year)/m$^2$ (SE), respectively. A common pattern is noticed in the

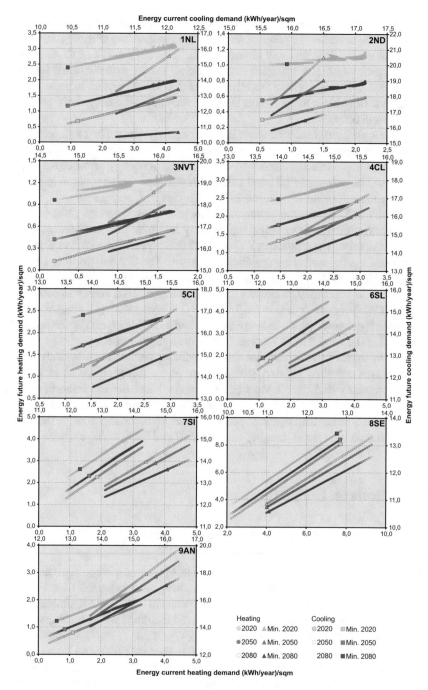

**Fig. 3.5**  Current energy heating and cooling demand versus future energy heating and cooling demand for 2020, 2050, 2080 under the A2 (medium-high) emissions scenario in the 9 climate zones of Chile

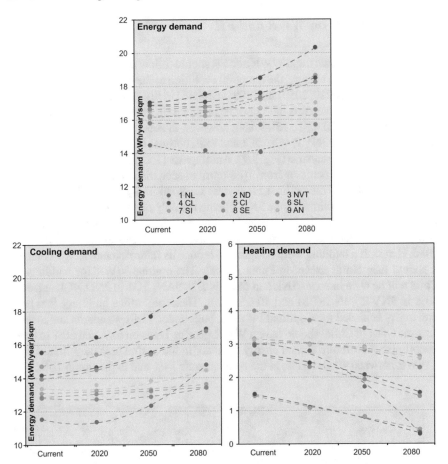

**Fig. 3.6** Evolution of energy demand for the optimal combination of WWR and FR for 2020, 2050, 2080 under the A2 (medium-high) emissions scenario in the 9 climate zones of Chile

evolution of the cooling and heating demand of the minimum annual energy demand cases, that is, an increase in cooling (0.53–4.47 (kWh/year)/m$^2$) and a decrease in heating demand (0.54–2.62 (kWh/year)/m$^2$). Zones 6SL, 7SI and 8SE practically balance out these increases and decreases and their annual energy demands remain constant (0.02–0.27 (kWh/year)/m$^2$). Whereas in zones 2ND, 3NVT, 4CL, 5CI and 9AN the increment of cooling demands produces an increment in the annual energy demand, despite the fact that heating demands decrease (0.48–3.27 (kWh/year)/m$^2$). A particular case study occurs in zone 1NL, in which the optimized case abruptly increases the cooling demand (3.25 (kWh/year)/m$^2$) and also noticeably decreases the heating demand (2.62 (kWh/year)/m$^2$).

### 3.4.2  Evolution on WWR and FR

In order to clarify which trend WWR and FR should follow in the future to counteract
the effect of a climate change, the temporal evolution of the optimized WWR and FR
in the present time, 2020, 2050 and 2080 is depicted in Fig. 3.6. Regarding the WWR,
there are locations where this ratio must be increased in order to lower annual energy
demand in 2080 (1NL, 6SL, 7SI, 9AN); however, in zone 2ND, this ratio should be
reduced. In locations 3NVT, 4CL, 5CI and 8SE, optimal values of WWR are similar
for every considered scenario (Fig. 3.7). Antofagasta (1NL) is the location where the
optimal WWR ratio increases from 10% (current scenario) to 60% in 2080.

In most locations, the optimal FR figures for 2080 imply that the building must
face North, with the climate zone 3NVT being the only one where the same optimal
FR is kept constant for the four scenarios (Fig. 3.7). In zone 8SE, North orientation
is not recommended in 2080, as the FR evolves from 1.250 to 2.178 in the year
2080. Hence, if a building were designed to reduce its future annual energy demand,
it should face North except in Punta Arenas. The optimal WWR for buildings in
Chile will be 60% in zones 1NL and 6SL; 54% in 9AN; 51% in 2ND, 4CL and 5CI;
40% in 3NVT; 43% in 7SI and 10% in 8SE. Therefore, office buildings that have
previously achieved their optimal FR ratio, built under TDRe standards and located
near the coast of Chile, should have a WWR ranging between 51% and 60% in order
to optimize their future annual energy demand; in the central area they should have
a WWR between 40 and 51%, in the Andean zone 54% and in the South 10%.

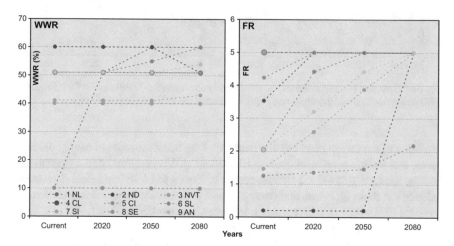

**Fig. 3.7**  Optimal WWR and FR values for 2020, 2050, 2080 under the A2 (medium-high) emissions
scenario in the 9 climate zones of Chile

## 3.5 Discussions

It has been proved that climate change has a significant impact on energy performance of office buildings in all climate contexts of Chile, as well as on the features of their basic design, such as FR and WWR. The optimization of the FR and WWR in the early design stages, taking into account a climate that would become warmer and the current Chilean building standard (TDRe), could result in a decrease in their annual energy demand. This will help in improving the optimization process of this kind of buildings, so that they will be more resilient during their lifetime. When considering this matter, using this study, the following outcomes can be outlined:

With respect to the relation between climate variables and energy demand for buildings, the following facts have been clarified in this paper. Predicted tendencies for 2020, 2050 and 2080 foresee a Chilean scenario where the climate will become hotter and drier for all locations; solar radiation levels will basically remain constant. With similar levels of GHR, lower RH and higher DBT, buildings will face a climatic scenario where cooling energy demand for air conditioning will rise. As the three scenarios are not very distant from present times (2, 32 and 62 years) and bearing in mind that they could embrace the expected lifespan of buildings designed and built today, these facts should be taken into account in the design of present buildings.

The influence of different FRs can be understood both from their evolution through time, and their tendency for a fixed time scenario. In regards to the first one, given a fixed location and a fixed scenario (2020, 2050 and 2080), the energy consumption seems to be quite independent from the FR variable, as little variance can be observed. This means that shape and orientation of the building is not strongly related with energy consumption for a given scenario in a given location. However, if the evolution of this variable is observed through time, it can be noted that the FR that gives the minimum energy consumption for all location tends to stabilize at around 5, except for location 8SE, where it is around 2. Thus, when designing buildings bearing in mind future climate scenarios, it should be taken into account that their FR should range around these reference values.

The influence of WWR shows, however, a different tendency. If location and climate scenario (2020, 2050, 2080) are fixed, expected energy consumption is strongly correlated to WWR. However, when its evolution is observed through the three considered scenarios, their reference values for minimum energy consumption do not change or only change slightly (50, 60%), except for the 1NL location, where an abrupt change is seen from 10 to 60%. Hence, buildings that should be adapted to this climate scenario should adopt a fixed WWR for each location, but careful attention should be paid, as a slight variation in this ratio would greatly affect energy demand.

If these two variables are combined (FR and WWR) and a batch test is performed for each location in order to devise the tendency for the energy consumption, the general tendency shows that, even when trying to find the minimum for each scenario (2020, 2050, 2080), energy consumption will inevitably rise, except for the locations 6SL, 7SI and 8SE. However, that decrease can be considered as non-significant, compared with the increase observed in the rest of locations. For some locations, the trend shows nearly a linear distribution, whereas in others it adopts the distribution of a point cloud, and that means that in some locations the tendency can be identified more easily, whereas in others is not so clear what may happen in future scenarios.

As this chapter was aimed at predicting the evolution of energy demand in future scenarios of climate change (2020, 2050, 2080), setting two variables as free (WWR and FR), and leaving the rest as fixed, the main conclusion that can be drawn from the study is this. No matter how much WWR and FR are optimized, energy demand of buildings will increase under the considered scenarios. This does not imply a disregard of these variables in the design process when trying to achieve a low energy design, as they are a commonplace in the basics of passive design. The main point is that under the considered evolution of climatic variables, the optimization of these variables will only be able to contain building's energy demand to skyrocket, but they will not be able to diminish it.

As a consequence of the former, with the objective being to reduce energy demand under future climate scenarios, other variables, which have been set as fixed in this study, such as improved constructive systems or changes in the internal heat loads profile, should be parameterized and set as free, in order to check their influence on future energy demand. Considering specific boundary conditions for sundry local environments that focus on real case studies, rather than on abstract prototypes, is also crucial to clarify the forecasted evolution of energy demand. Therefore, further study on this matter is necessary in order to devise effective strategies, which will help in reducing energy demand of buildings in a context of a changing climate.

# Chapter 4
# Multiple Linear Regressions

## 4.1 Introduction

This chapter intends to develop a mathematical model that allows predicting, with an acceptable degree of uncertainty, the energy consumption and $CO_2$ emissions for the office buildings in Chile. Through the multivariable regression method, diverse equations will be produced that will bear in mind the parameters mentioned for the different locations. In this way, the designers will be able to know the consequences that their decisions will have on the energy consumption and $CO_2$ emissions. This research has an eminently practical nature and is susceptible to being applied in the future design and construction of buildings.

The research sets out a simple but realistic approach to predict the energy consumption and the $CO_2$eq emissions of a standard office building located in different cities of Chile, based on the location, total surface area, number of stories, orientation, window-to-wall ratio, efficiency of the systems and emission factors (Fig. 4.1). ISO 13790:2008 was used to build and simulate individual setups of the building which were generated using national standards. 77,000 simulations were made for each location to create a group of integrated data that covers the complete range of design parameters. Below, the results of the energy simulations were put into practice in a regression equation system to predict the energy consumption and the emissions under each design variables scenario.

The number of floor levels varies between 1 and 5. The surface area of the floor varies between 500 and 2000 m$^2$, in steps of 100 m$^2$. The form ratio is defined as the proportion between the length of the face orientated towards the North and that orientated to the East. The buildings are assumed to be perfectly orientated as per a N-S axis. The values of this parameter vary between 5 (an extended building in a E-W direction) and 0.2 (extended in a N-S direction), varying the building's degree of compactness, in this way, in stages of 0.2. The glazing percentage is considered to be uniform for all the facades and varies between the values permitted by the standard, which are 10 and 60%; the variation has been established in 5% intervals.

Given that the choice of a determined piece of air-conditioning equipment is the decision of the designer, and that this is also influenced by numerous factors (client preferences, size of the building, project budget, etc.), a random value of COP, EER,

© The Author(s) 2018
C. Rubio-Bellido et al., *Energy Optimization and Prediction in Office Buildings*,
SpringerBriefs in Energy, https://doi.org/10.1007/978-3-319-90146-6_4

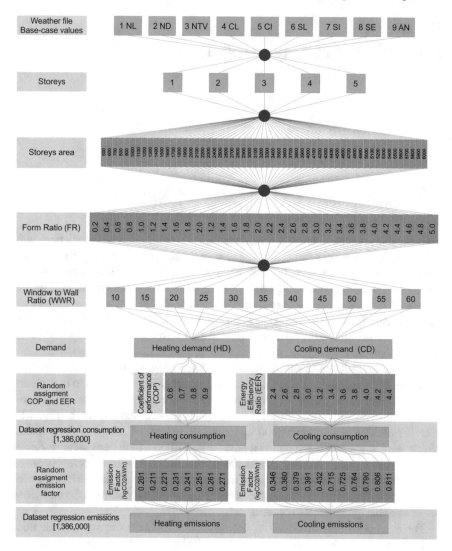

**Fig. 4.1**  Parameter tree

HEF and CEF has been assigned for each one of the 77,000 available cases, that sits between the maximum and minimum limits established in Table 4.1 (Fig. 4.1).

In this way, the summarized calculation process is as follows. For a given location, following the ISO 13790:2008 standard, a heating and cooling demand is generated for 77,000 possible office buildings, resulting from the multiple combinations of the following variables, number of stories, floor area, shape coefficient and glazing percentage of the facades. Once the heating and cooling demands are obtained, each building is assigned air-conditioning equipment whose energy efficiency is dis-

**Table 4.1** Ranges and intervals for geometry and HVAC calculation procedure

| Input parameters | Unit | Range | Interval | Number of items[a] |
|---|---|---|---|---|
| Number of stories (NS) | Nos. | 1–5 | 1 | 5 |
| Floor area (FA) | $m^2$ | 500–6000 | 100 | 56 |
| Form ratio (FR) | – | 0.2–5 | 0.2 | 25 |
| Window-to-wall ratio (WWR) | % | 10–60 | 5 | 11 |
| Coefficient of performance (COP) | % | 60–90 | 10 | 4 |
| Energy efficiency ratio (EER) | – | 2.5–4.4 | 0.1 | 15 |
| Heating emission factors (HEF) | $(TCO_2eq/kWh)$ | 0.201–0.271 | 0.01 | 8 |
| Cooling emission factors (CEF) | $(TCO_2eq/kWh)$ | 0.346–0.811 | 0.01 | 11 |

[a]The combination of the items generates 77,000 heating demands and other refrigeration demands for each one of the locations studied

tributed randomly, providing 77,000 heating energy consumption options and 77,000 cooling consumption options. After that, in order to transform electrical consumption into $CO_2eq$ emissions, a random $CO_2eq$ emission factor is assigned from Table 4.1, and depending on the electrical system to which each prototype should be connected to, the $CO_2$ emission factor is assigned to these 154,000 options. Finally, multiplying by 9 locations, we obtain 1,386,000 possible cases, which constitute the two databases used to prepare the multivariable regression models (Fig. 4.1).

## 4.2 Energy Consumption

Regression models have been devised for the calculation of the energy consumption (EC) for each one of the 9 climatic zones already defined. The model contains coefficients that represents, as close as possible, the distribution of the data; each model has one constant term ($\beta_0$) and coefficients ($\beta_p$) that affect the 6 independent variables, following in this way the form already seen in this equations.

$$Y = \beta_0 + \beta_1 X_1 + \cdots + \beta_p X_p + \varepsilon \tag{4.1}$$

For each of the climatic zone, the equation that best represents the energy consumption is given below. Following this common pattern, the equations for each one of the considered climatic zones are as follows (Eqs. 4.1–4.9):

Zone 1NL

$$EC = 10.6054 + \frac{1}{FA} 760.409 + E^{NS} 6.6294 \times 10^{-3} + \frac{1}{FR} 7.4437 \times 10^{-3}$$
$$- \frac{1}{WWR} 4.5996 + \ln(COP) \cdot 3.21246 + \ln(EER)4.16894 \quad (4.2)$$

Zone 2ND

$$EC = 14.6288 - \frac{1}{FA} 34.341 + E^{NS} 5.44267 \times 10^{-4} + \frac{1}{WWR} 1.12755$$
$$- \ln(COP) \cdot 1.13665 - \ln(EER) .69142 \quad (4.3)$$

Zone 3NVT

$$EC = 12.2494 - \frac{1}{FA} 48.7499 + E^{NS} 6.70744 \times 10^{-3} + \frac{1}{FR} 1.09518 \times 10^{-3}$$
$$+ \frac{1}{WWR} 0.898009 - \ln(COP) \cdot 0.378144 - \ln(EER) \cdot 5.49944 \quad (4.4)$$

Zone 4CL

$$EC = 12.9473 + \frac{1}{FA} 320.757 + E^{NS} 1.95948 \times 10^{-3} + \frac{1}{FR} 4.05568 \times 10^{-3}$$
$$- \frac{1}{WWR} 2.08727 - \ln(COP) \cdot 1.28408 - \ln(EER) \cdot 5.1702 \quad (4.5)$$

Zone 5CI

$$EC = 12.6872 + \frac{1}{FA} 286.71 - E^{NS} 3.64316 \times 10^{-5}$$
$$- \frac{1}{WWR} 2.09835 - \ln(COP) \cdot 1.11644 - \ln(EER) \cdot 5.19887 \quad (4.6)$$

Zone 6SL

$$EC = 12.0439 + \frac{1}{FA} 509.268 + E^{NS} 4.08027 \times 10^{-3} + \frac{1}{FR} 8.65602 \times 10^{-3}$$
$$- \frac{1}{WWR} 4.12168 - \ln(COP) \cdot 1.2269 - \ln(EER) \cdot 4.71123 \quad (4.7)$$

Zone 7SI

$$EC = 12.3528 + \frac{1}{FA} 708.38 + E^{NS} 2.75877 \times 10^{-3} + \frac{1}{FR} 9.26539 \times 10^{-3}$$
$$- \frac{1}{WWR} 6.48265 - \ln(COP) \cdot 1.33644 - \ln(EER) \cdot 4.76629 \quad (4.8)$$

Zone 8SE

$$EC = 12.353 + \frac{1}{FA}1580.01 + E^{NS}13.4577 \times 10^{-3} + \frac{1}{FR}22.3996 \times 10^{-3}$$
$$- \frac{1}{WWR}19.7227 - \ln(COP) \cdot 6.41663 - \ln(EER) \cdot 4.2491 \qquad (4.9)$$

Zone 9AN

$$EC = 12.6116 + \frac{1}{FA}573.704 + E^{NS}38.6045 \times 10^{-3}$$
$$- \frac{1}{WWR}5.15063 - \ln(COP) \cdot 1.15101 - \ln(EER) \cdot 5.06677 \qquad (4.10)$$

All of the nine equations adopt the same form with some remarks. The term 1/FR has been deleted from zones 2, 5 and 9 after conducting a p-value analysis, which determined that the parameter FR has no influence in the energy consumption in those climates. This is not minor feat when trying to devise a sort of mathematical model that finds adaptation to a variety of climates with remarkable differences, which is the case of Chile. In this occasion, the decision was to remove the term for the sake of simplicity; however, only 3 out of 54 constant terms were removed, which indicates that the model, in its actual form, represents with enough accuracy the variation of climates in Chile as per OGUC and more recently the non-compulsory standard TDRe. Anyway, precaution should be exercised in informing the potential users of which variables are not significant in order to reduce the computational load.

Other important issue when devising models for the early stages of design is to bear in mind the real accuracy of the model and the expected error, which can be measured, respectively, using the coefficient of determination $R^2$, the Standard Error (SE) and the Mean Absolute Error (MAE). Table 4.2 shows these parameters for the 9 considered climate zones. Following the common rule of statistics, when $R^2$ values are above 95% the model should be considered reliable enough. In this case, the model for zones 1, 7 and 8 are between 90 and 95%, which can be considered fair but not ideal. The rest of the 6 models are all above 95%. For each one of them, the SE and the MAE will give information about the expected inaccuracy of a given calculation.

This concept can be furtherly clarified using a concrete example. Let's consider that an office building is planned to be built in the city of Concepción, located in zone 6SL. The building will have a floor area of 1500 m$^2$, distributed in 3 floors; the plan will adopt the shape of a square oriented towards North-South axis, that is, form ratio (FR) will be the unit; the glazing percentage will be 30%, that is, a WWR of 30; finally, the considered HVAC equipment will be a multi-split system with reversible heat pump (heating and cooling) with a COP of 80 and a EER of 3.3. Just by substituting all terms into the aforementioned equation for zone 6SL we can have an estimation of the energy consumption as 4.160 KWh/m$^2$. Considering also the MAE of the model, this statement should be reformulated in this way.

**Table 4.2** Regression coefficients and model precision (energy consumption)

| Code | 1NL | 2ND | 3NVT | 4CL | 5CI | 6SL | 7SI | 8SE | 9AN |
|------|------|------|------|------|------|------|------|------|------|
| $R^2$ | 93.54 | 96.79 | 98.05 | 96.94 | 97.58 | 95.26 | 93.53 | 91.81 | 96.10 |
| SE | 0.27 | 0.22 | 0.15 | 0.26 | 0.23 | 0.30 | 0.34 | 0.54 | 0.31 |
| MAE | 0.20 | 0.18 | 0.11 | 0.20 | 0.18 | 0.23 | 0.27 | 0.41 | 0.23 |

The predicted EC should fall between $4.160 \pm 0.11$ kWh/m$^2$. Going into absolute figures, the energy consumption of the whole building should be in the range of $6240 \pm 165$ kWh. Of course, spreadsheet software may be useful when doing these calculations as the coefficients can be tabulated; in that way, calculation become much more accessible to those who even do not have a deep understanding of energy simulations in buildings.

Also, some remarks can be made about the physical meaning of the model by analyzing the terms of the said equations, bearing in mind that these data should be handled carefully. First, the constant term ranges from 10.6054 (Zone 1NL) to 14.6288 (Zone 2ND), being around 12 in the rest of zones. This indicated that the minimum theoretical energy consumption that a designer could expect, in the best case scenario, would be around these figures (10–14 kWh/m$^2$). Floor area (FA), number of stories (NS) and Form ratio (FR) shall be understood together, because they would have no physical meaning alone (they are modified by inverse fractions and exponential functions and their sign swings from positive to negative depending on the climate zone). The three of them, understood as a whole, are related with the form of the building and therefore, with the so-called form factor; the equation could have included this parameter instead, but these three parameters allows for more flexibility because no previous calculations are necessary. The window to wall ratio (WWR) would suggest, at first glance, that the lower the WWR, the lower the energy consumption, as windows have always lower U values that the opaque envelope according to data from Chilean standard. So that in every zone this factor should be preceded by a positive sign, so lower WWR should yield smaller increments of energy consumption. But that relation is not that simple, so in this case the mathematical model fits WWR into an inverse fraction, affected by a constant term and a change in the preceding sign, depending on the climate zone. As an example, in cold climates, properly oriented windows can help with solar gains, and in mild climates the infiltration rates of windows can counteract the internal gains, which are particularly important in tertiary buildings. Last, the terms associated with COP and EER behave in a logic manner; all of them are preceded by negative signs and affected by a coefficient depending on the climate zone. In this case, a direct physical meaning can be inferred, because higher COP and EER, that is, better heating and cooling equipment, would mean a reduction in energy consumption.

This exercise should be performed after obtaining the multivariable regression model in order to gain a better grasp about the actual meaning of the whole equation, as well as each term itself. As previously stated, the questions that the designer should ask him/herself when analyzing each term is ¿has the term physical meaning by itself? The affirmative answer would correspond to COP and EER in this case. Second question: If it has no physical meaning by itself ¿is it related to various factors that could explain this lack of sense? That would be the case of WWR. Third question. If it has no physical meaning by itself ¿should it be understood together with other terms in order to make sense? That is the case of FA, NS and FR, which, all together, define the form of a given office building.

## 4.3  $CO_2$ Emissions

Once having the energy consumption of the office building, the same method can be used to estimate their $CO_2$ emissions, but introducing a new factor: The $CO_2$ equivalent emission factor ($CO_2$eq). This factor is related to the way electricity is generated in a given country or region, and inform the designer on how much $CO_2$ is produced to generate 1 kWh of electric energy that will be, at last, used in the building. Two assumptions are used. First, the HVAC systems runs exclusively on electricity; second, the $CO_2$eq is independent of the building features, and is extracted from statistical data from the electric grid. Energy companies usually provide with this data, which actually varies year to year, depending on the energy mix. In the case that the building should be prone to use other source of energy (for example, diesel or natural gas boilers for heating) another factor should be used, $CO_2$eq (Kg/Ton). In this case, two new terms have been introduced into the equation: The heating emission factor (HEF) and the cooling emission factor (CEF). Taking into account these considerations, the regression equations for each zone are as follows (Eqs. 4.11–4.19).

Zone 1NL

$$CO_{2eq} = e^{\begin{array}{l} 0.750595 + \frac{1}{FA}41.8966 - E^{NS}1.34961 \times 10^{-4} - \frac{1}{WWR}0.225368- \\ \ln(\text{COP}) \cdot 0.285616 - \ln(EER)\} \cdot 0.7302 + HEF \cdot 1.16794 + CEF \cdot 1.32949 \end{array}} \quad (4.11)$$

Zone 2ND

$$CO_{2eq} = e^{\begin{array}{l} 1.29841 - \frac{1}{FA}2.85706 + E^{NS}2.66382 \times 10^{-4} - \frac{1}{FR}4.81831 \times 10^{-4} + \frac{1}{WWR}0.24591 - \\ \ln(\text{COP})\} \cdot 0.124535 - \ln(EER) \cdot 0.947335 + HEF \cdot 0.284265 + CEF \cdot 1.53482 \end{array}}$$

$$(4.12)$$

Zone 3NVT

$CO_{2eq}$

$$= e^{\begin{array}{l} 1.18476 - \frac{1}{FA}10.4812 + E^{NS}1.35099 \times 10^{-3} - \frac{1}{FR}1.91223 \times 10^{-4} + \frac{1}{WWR}0.160358 - \\ \ln(\text{COP}) \cdot 0.0551293 - \ln(EER)\}cdot0.960037 + HEF \cdot 0.0640181 + CEF \cdot 1.76345 \end{array}}$$

$$(4.13)$$

Zone 4CL

$$CO_{2eq} = e^{\begin{array}{l} 1.14686 + \frac{1}{FA}18.7367 + E^{NS}6.31546 \times 10^{-5} - \frac{1}{WWR}0.0418683 - \\ \ln(\text{COP}) \cdot 0.150913 - \ln(EER)\}cdot0.830998 + HEF \cdot 0.427063 + CEF \cdot 1.42374 \end{array}}$$

$$(4.14)$$

### Zone 5CI

$$CO_{2eq} = e^{\begin{array}{l} 3.52613 + \frac{1}{FA}39.3627 - E^{NS}3.58186 \times 10^{-4} - \frac{1}{WWR}0.0115162 - \\ \ln(COP) \cdot 0.309527 - \ln(EER)\}cdot2.89103 + HEF \cdot 0.817204 + CEF \cdot 4.89869 \end{array}}$$

$$(4.15)$$

### Zone 6SL

$$CO_{2eq} = e^{\begin{array}{l} 1.0351 + \frac{1}{FA}26.066 - E^{NS}9.75502 \times 10^{-5} - \frac{1}{WWR}0.148346 - \\ \ln(COP) \cdot 0.16862 - \ln(EER) \cdot 0.786633 + HEF \cdot 0.520632 + CEF \cdot 1.37929 \end{array}}$$

$$(4.16)$$

### Zone 7SI

$$CO_{2eq}$$
$$= e^{\begin{array}{l} 1.02811 + \frac{1}{FA}36.7475 - E^{NS}2.77123 \times 10^{-4} + \frac{1}{FR}5.41478 \times 10^{-4} - \frac{1}{WWR}0.304362 - \\ \ln(COP) \cdot 0.188558 - \ln(EER) \cdot 0.767077 + HEF \cdot 0.599526 + CEF \cdot 1.3633 \end{array}}$$

$$(4.17)$$

### Zone 8SE

$$CO_{2eq}$$
$$= e^{\begin{array}{l} 0.774452 + \frac{1}{FA}80.7924 + E^{NS}5.31195 \times 10^{-4} + \frac{1}{FR}1.1994 \times 10^{-3} - \frac{1}{WWR}0.971469 - \\ \ln(COP)\} \cdot 0.415091 - \ln(EER) \cdot 0.600031 + HEF \cdot 1.73368 + CEF \cdot 1.09288 \end{array}}$$

$$(4.18)$$

### Zone 9AN

$$CO_{2eq}$$
$$= e^{\begin{array}{l} 1.12736 + \frac{1}{FA}30.0675 - E^{NS}7.26357 \times 10^{-4} + \frac{1}{FR}4.27955 \times 10^{-4} - \frac{1}{WWR}0.159419 - \\ \ln(COP) \cdot 0.151394 - \ln(EER) \cdot 0.830492 + HEF \cdot 0.443889 + CEF \cdot 1.40514 \end{array}}$$

$$(4.19)$$

Follow the same pattern, focusing on the physical differences between both regression models, the following remarks shall be made. Firstly, one might think that CO$_2$eq emissions could be obtained just by multiplying the result for EC by the HEF or the CEF, on a case basis. However, the procedure bears no so simple relation. As both heating and cooling are introduced in the equation for EC, the equivalent emission factors (HEF and CEF) cannot be introduced in such a direct way, and that is why another equation becomes necessary.

In the same fashion, after conducting a *p*-value analysis, it can be concluded that the FR is not statistically relevant when predicting CO$_2$eq emission in zones 1NL,

4CL, 5CI and 6SL; besides, the WWR is not necessary in zone 5CI. Out of 72 terms, 5 have been found to be irrelevant; therefore, the chosen variables seem to be adequate, though some simplifications can be made, especially in zone 5CI, in order to alleviate the computational load.

The regression coefficients (Table 4.3) indicate that this model is even more accurate that the one for EC. All zones have $R^2$ values above 95%; SE and MAE should be used, again, to have an idea of the inaccuracy of the model.

Following the same example that in the previous section, let's consider that an office building is located in Concepción (zone 6SL). A random value for HEF of 0.210, and for CEF of 0.3709 are assigned. Using the proposed equation, the expected $CO_2$eq emission per square meter would be $1.57 \pm 0.02$ $kgCO_2/m^2$; going into absolute figures, that would result in $2361 \pm 30$ $kgCO_2$. Roughly speaking, the said building would contribute with 2.3 tons of $CO_2$ due to the energy consumption in heating and cooling.

In this occasion, the physical meaning of the model is not so clear because the whole equation is affected by an exponential function; sometimes this time of mathematical operations become necessary in order to improve the accuracy. The statistical software used for devising the regression model will perform these operations automatically, trying to balance the most accurate model with the simplest expression. Anyway, some check would reveal whether the model has some mistake. COP and EER are all affected by a negative sign in all zones, which seems logical; besides, HEF and CEF coefficients are all preceded by positive signs, which indicates that, the higher $CO_2$ emissions from the electric grid, the higher the $CO_2$ emissions from the building that is connected to that grid. The rest of the terms have changing signs and very different coefficients because, as said before, they have a more complex interplay that do not allow for a direct interpretation.

## 4.4   Regression Models Validation

The $R^2$, SE and MAE give information about the accuracy and consistency of the mathematical expression itself. However, in order to actually check whether that expression actually reflect the phenomenon that is trying to reproduce (a physical phenomenon, a calculation procedure, etc.) it is necessary to undertake additional checks.

In this case, the objective of the multivariable regression model is to reproduce in a more simplified manner the calculation procedure per ISO 13790:2008, which can be considered, in this occasion, the base case. That is why, amongst all the available data, 75% was used for generating the regression equations, and the remaining 25% (training data) is used to check the results against the base case. It is important, in such a manner, to have an ample database and to store this remaining 25% for additional checking; of course, more data means more accuracy. In this case, 1,386,000 cases were considered, amongst 1,039,500 were used to generate the 18

**Table 4.3**  Regression coefficients and model precision ($CO_2$eq emissions)

| Code | 1NL | 2ND | 3NVT | 4CL | 5CI | 6SL | 7SI | 8SE | 9AN |
|------|-----|-----|------|-----|-----|-----|-----|-----|-----|
| $R^2$ | 98.82 | 99.37 | 99.56 | 99.29 | 96.83 | 98.87 | 98.73 | 96.89 | 99.01 |
| SE | 0.03 | 0.03 | 0.03 | 0.03 | 0.22 | 0.04 | 0.04 | 0.05 | 0.04 |
| MAE | 0.02 | 0.02 | 0.02 | 0.02 | 0.18 | 0.03 | 0.03 | 0.04 | 0.03 |

regression equations and 346,500 to check the accuracy against direct results per ISO 13790:2008.

At first, results from both the regression model and the ISO procedure fits pretty well, because the cloud point is distributed uniformly around the $X = Y$ line (Fig. 4.2). It is also representative that the point clouds for all zones are more compact for buildings with low energy consumption and more disperse when the consumption reaches higher figures. The minimum energy consumption for all zones ranges around 3.5–4–5 kWh/m², which contradicts the statement from Sect. 4.1; that is why, when trying to grasp the physical meaning of a multivariable regression model, caution should be exercised before checking results against training data. In such a similar manner, the maximum expected consumption should range between 8 and 18 kWh/m². This is an additional advantage of such verifications; with such amount of data, minimum and maximum values can be established and therefore outliers can be easily identified; for example, in the event that energy simulations for an office building located in zone 5CI were undertaken, as long as the building features fall into the ranges of this regression model, unitary energy consumption should be between 3.5 and 10.5 kWh/m². Different values should be deemed, at first, as outliers and calculations should be verified again.

As commonplace in statistics and multivariable regression models, residuals are present because the model is not 100% accurate. The question is how to analyze those residuals and how to determine if these residuals jeopardize the validity of such model. This questions can be answered by means of the studentized residual, which is a way of determining if an outlier is exerting a strong influence on the model in such a way that the regression equation can be altered. The basic idea consists on deleting a given observation (result) of the model and see how the regression model is refitted with the remaining n-1 observations. Then the results from the model with and without the given observation are compared, producing a deleted residual; as this deleted residual is dependent on the measurement unit, it has to be standardized, and that is why the deleted residual is divided by an estimate of its standard deviation, and that is called the "studentized residual". Logically, if the event that the model would be perfect, all studentized residuals would be zero; in statistics, it is commonly accepted that residuals over 3 are deemed as outliers. It is also important to note that the shape of the graph is deemed important; no patterns should be present if the model fits well and if the point cloud adopt a concrete shape (the most common are parabolic and funnel shape) there may be non-linearity in the data or heteroscedasticity, which are both undesirable effects. At this point, if deeper explanations are needed, specialized texts on statistics should be consulted.

In this case, studentized residuals are presented in Fig. 4.3. First, studentized residuals higher than 3 are present, despite their number is residual when compared with the whole cloud point; that is, the densest cloud is enclosed by two horizontal lines, +3 and −3. The presence of outliers in this case can be explained by the elevated amount of data, which may produce particular combinations of variables that would make no sense. Let's consider, as an example, an office building located in Punta Arenas, with a predominance of heating and no presence of cooling. The designer makes the dumbest choice, a heating system with the lowest possible COP

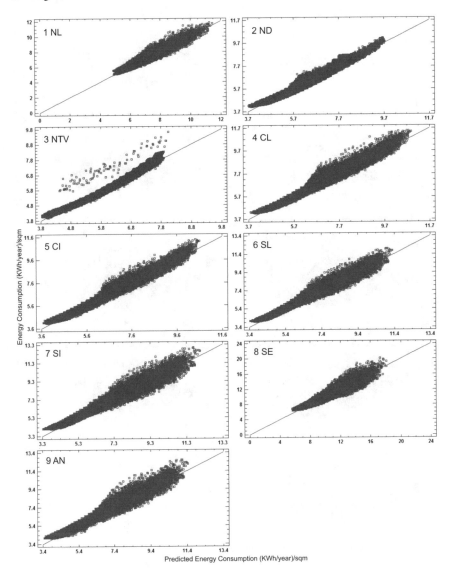

**Fig. 4.2** Predicted energy consumption versus ISO 13790:2008 energy consumption

and a cooling system with the highest available EER, unless it would be unnecessary. Besides, a poor choice of the considered design variables (predictors) could lead to unrealistic results (in this case, worst orientation possible, a high WWR, a poor form factor, etc.). The point is, when such elevated amount of data is treated, outliers will be present, but their number should be marginal when compared to the total number of points, as in this case.

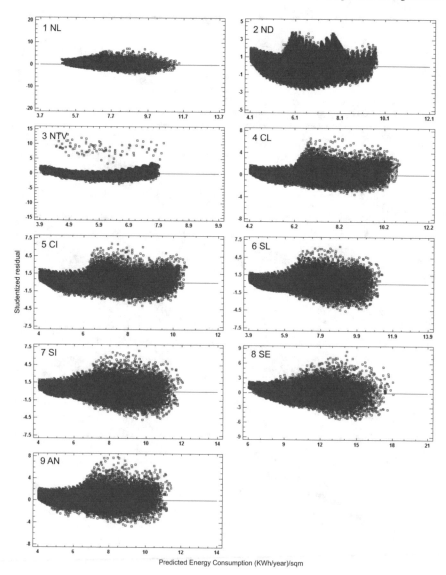

**Fig. 4.3**  Studentized residual energy consumption

The shape of the cloud point, as mentioned before, is also deemed important. In this case, the cloud points are distributed randomly around the horizontal line $Y = 0$, which is a good indicator; that means that the residuals do not follow a particular trend; that is, they bear no mathematical relation with the regression equation. Besides, it can be observed that residuals larger than 3 are more numerous for higher energy consumptions. This means that the model losses accuracy until some extent when trying to predict higher energy consumptions, unless it can be still considered reliable.

Some remarks should be make regarding zone 3NVT; their graphs show a particular shape, because both fitting between ISO and regression consumptions and the studentized residual show a group of points detached from the main group. This phenomenon can be explained by the fact that energy demand for heating is below 10%, but very close. As stated in Sect. 2.2.5, if heating demand is assumed, in this case, by cooling equipment in heat pump mode, the results can be distorted. ISO calculation procedure does not make this simplification, so that results will not fit and, therefore, the studentized residual will be outstanding in those few cases.

This checking has also been performed for $CO_2$ equivalent emissions (Fig. 4.4). The fitting between the calculation per ISO 13790:2008 and per the multivariable regression model show that the point clouds are pretty much compact. The equations for this model are all affected by an exponential sign and that is why the model fitting seems better. Talking about figures, $CO_2$ equivalent emissions are in the range of 0.4–2.2 $kgCO_2eq$/year $m^2$ for the whole country. Resuming the former example, a building located in zone 3NVT with an estimated $CO_2$ equivalent emissions of 1.57 $kgCO_2eq$/year $m^2$ would be located in the upper tier of the graph. Outliers or abnormal values for $CO_2$ equivalent emissions can be also be easily identified by virtue of this analysis.

The analysis for the studentized residual of $CO_2eq$ emissions (Fig. 4.5) also shows that, in general terms, no clear patter can be detected, unless for zone 5CI, where the point cloud seems to have a paraboloid shape. In all cases, residuals higher than 3 are present, but, once more, this can be considered admissible due to the elevated amount of data and the presence of such numerous predictor variables, which make possible to have "unreal" design combinations that will produce, at last, unreal figures for $CO_2eq$ emissions. In all zones, it can be noted that small clusters are formed around certain values of X axis. Precisely, 11 clusters can be identified in each zone, especially in zones 2ND, 3NVT and 4CL, which would correspond to the 11 assigned CEF. This suggest the strong influence that any variation in the CEF exerts on the model. Taking each cluster separately, the residuals have a perfect random distribution around zero.

The paraboloid shape in zone 5CI suggest that there is some room for improvement in the model; taking each one of the 11 clusters that can be identified, the residuals have a random distribution, but taking them as a whole, a pattern can be identified. In this case, the model would have some room for improvement (for example, transforming some variable). Anyway, the accuracy of the model seems balanced for all locations so, in order to keep the coherence, in this case no further transformations will be necessary.

That is que question that the designer should bear in mind when devising a large number of multivariable regression models: Keeping a common pattern for all of them may compromise the accuracy of some; in this case, the exponential function gave a hint about the best balance between accuracy and applicability of such models.

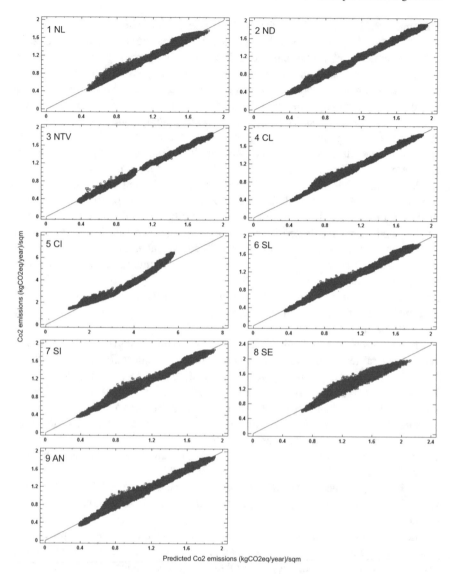

**Fig. 4.4**   Predicted energy consumption versus ISO 13790:2008. $CO_2$ emissions

## 4.5   Discussions

Mutivariable regression is a technique widely used in many fields of science. It also finds wide application in architecture and engineering, and in this case it has been used to estimate two figures that are of interest when dealing with low energy design: Energy consumption and $CO_2$ emissions.

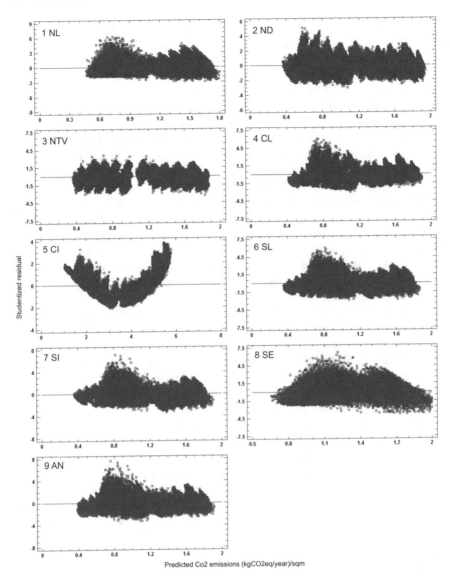

**Fig. 4.5** Studentized residual. $CO_2$ emissions

This methodology can be replicated for any kind of building (residential, educational, administrative…) in their early stages of design, if the following conditions are met. First, the designer should have an educated guess about the design features of the building itself; that will give a hint about the predictor variables. In this case, office buildings were found to adopt a rectangular shape, with different orientations and percentages of glass. More complex shapes (L-shaped or H-shaped, for instance) or other variables could also be included only if the second condition is met: The predictor variables should be parametrized, which means that they should be prone to be defined by numbers and/or mathematical expressions. Third, a base case for a proper checking should be available; in this case, ISO 13790:2008 was used as the base case for checking the accuracy of the model.

These models are especially useful during the early stages of design, when important decisions about the geometry, orientation and façade design are made. Other variables, such as internal loads and U values of the external envelope can be fixed, as in this case, per building codes. In such way, they allow for trial and error of numerous combinations of the predictor variables and therefore the best combination of such parameters can be clarified before going deeper into executive design. Besides, fixing all the predictor variables except one will allow for a detailed study about the influence of such variable in the energy consumption and $CO_2$ emissions. In the specialized literature, this technique, often denoted as parametric design, has proven to be highly effective, because optimum values for those parameters that have a great impact in the design of the building can be clarified in the first stages of design. Therefore, these techniques are placed in the early stages of design, that is, basic design, feasibility studies or first drafts.

These models should be understood to estimate the energy consumption and $CO_2$ emissions of office building, but no to give exact figures; that is why the MAE should always be introduced to give a range (minimum and maximum expected values) of such parameters. These estimations may be handy in the first stages of design, when estimations are more useful than exact values. Drawing a parallelism, in this stage the exact dimensions of the structure of the building would not still be clarified, but an estimation should be sufficient.

Other application of the multivariate regression models is that they provide simple equations to represent a complex calculation procedure; in this way, they can also be used to enhance building policies or provide them with tools in order to check compliance with rating schemes, such as LEED or BREEAM.

The main limitations of the models hereby presented relies in the fact that the design features of the buildings are expressed as mathematical parameters, and therefore the design is parametrized and, until some extent, simplified. That is why these models find no application when dealing with personalized designs or any kind of particular buildings. Structures where a typology could be easily identified (offices, schools, hospitals, social dwellings...) are prone to be parametrized.

# Chapter 5
# Artificial Neural Networks

## 5.1 Introduction

This chapter intends to demonstrate the performance and reliability of ANN in predicting large scale data not only for a single parameter, but for three of them (energy consumption, energy demand and $CO_2$ emissions) in relation to a large-scale sample of buildings, with all the issues associated to them, such as the nonlinearity of problems related to building design and performance.

It is expected that the results of the study will be of help in two main areas. First, assisting designers in the early stages of design, being able to grasp the demand of resources of their design with a few variables. Second, the reference values expected from the ANN results will be of help in assisting future policy makers in establishing realistic goals in order to reduce, or at least contain, the energy demand, energy consumption and $CO_2$ emissions for these buildings.

Following the ISO 13790:2008 standard, a heating and cooling demand is generated for 77,000 possible office buildings in Santiago, resulting from the multiple combinations of the number of storeys (NS), floor area (FA), form ratio (FR) and window-to-wall ratio (WWR). Once the heating and cooling demands are obtained, each building is assigned air-conditioning equipment whose coefficient of performance (COP) and Energy efficiency ratio (EER) is distributed randomly, providing 77,000 heating energy consumption options and 77,000 cooling consumption options. After that, in order to transform electrical consumption into $CO_2$eq emissions, a random $CO_2$eq emission factor, that is, the $CO_2$ emission factor is assigned to these 154,000 options (Fig. 5.1). Summing up, the training data used in this study is composed of 77,000 possible designs for office buildings, formed by combining all the variables depicted in Fig. 5.1. These amounts of data are not composed by a single variable, but by a complex interplay of several design parameters that bear both linear and non-linear relations.

© The Author(s) 2018
C. Rubio-Bellido et al., *Energy Optimization and Prediction in Office Buildings*,
SpringerBriefs in Energy, https://doi.org/10.1007/978-3-319-90146-6_5

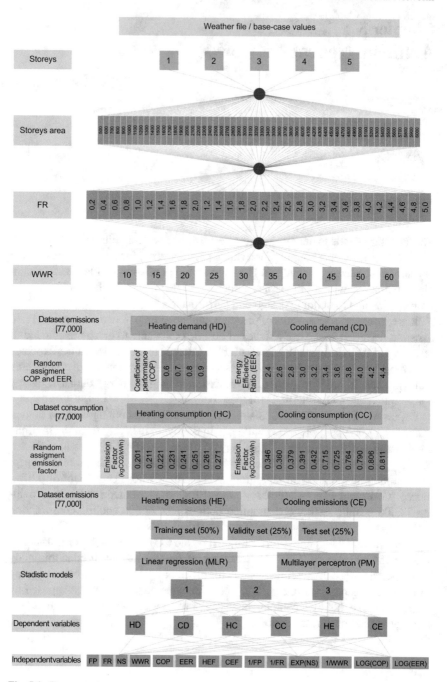

**Fig. 5.1** Parameter tree

**Table 5.1** Correlations matrix of the non-transformed predictor variables

|     | FP      | FR      | NS      | WWR     | COP     | EER     | HEF    |
|-----|---------|---------|---------|---------|---------|---------|--------|
| FR  | 0.0000  | –       |         |         |         |         |        |
| NS  | 0.0000  | 0.0000  | –       |         |         |         |        |
| WWR | 0.0000  | 0.0000  | 0.0000  | –       |         |         |        |
| COP | 0.2869  | −0.0437 | −0.5485 | −0.2060 | –       |         |        |
| EER | −0.0009 | 0.0000  | 0.0000  | 0.0000  | −0.0028 | –       |        |
| HEF | 0.2230  | −0.0349 | −0.4241 | −0.1596 | 0.7092  | −0.0321 | –      |
| CEF | 0.0001  | −0.0001 | 0.0000  | −0.0003 | 0.0022  | −0.0660 | 0.4769 |

## 5.2  Data Description

The predictor variables used for the regression models and multilayer perceptron for the calculation of the cooling and heating demands are solely those related with the building's geometry, FP, FR, NS and WWR (Table 5.1). The consumption models are formed through the geometry's predictor variables added to the air-conditioning systems' energy performance predictor variables (COP and EER). The emissions factors must be added to these variables to determine the heating and cooling emissions models (HEF and CEF).

It is seen, in Table 5.1, that most of the linear correlation coefficients between the non-transformed predictor variables are 0 or are very low, with the exception of the COP and HEF variables. The COP, EER, HER and CEF variables have been allocated randomly to the calculation process to determine the dependent variables to generate the models. The allocation has been done in MsEXCEL® through the RAND formula. This has been done because the COP only has 4 values to allocate and HEF 8, generating a linear correlation between said predictor variables and the rest.

Given that the NS variable presents a discrete set of values, from 1 to 5, a variance analysis has been made of one factor with 5 levels for one of the remaining variables. Figure 5.2 presents the box-and-whiskers graphical representation for each variable. COP and HEF values present some outliers due to the following reason. Despite the considered buildings usually having to address both heating and cooling demands, in some cases the consumption for heating is statistically non-significant due to the climatic context of Santiago, so it has been established as a criteria that when the heating consumption represents less than 10% of the total, this demand will be covered by the cooling system itself, which can run on heat pump operation mode. This criterion distorts the values for COP and HEF and explains the presence of some outliers.

As would be expected on examining these figures, only the analysis of variance (ANOVA) of the COP and HEF variables following the NS levels, was significant ($p$ value $< 0.001$). In the following section, the transformations made to the predicting variables are described to check if, in this way, it is possible to obtain more reliable regression models.

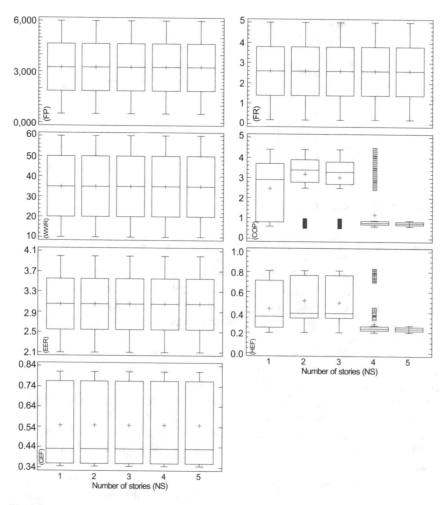

**Fig. 5.2** Box-and-whiskers FP, FR, WWR, COP, EER, HEF and CEF based on NS

## 5.3   Data Pre-processing

For a reasoned and reliable comparison of the different methods, the data set has been divided randomly into three parts, with respective sizes of 50, 25 and 25. The first part is used to adjust the models (38,500 cases), therefore acting as the training set. The second sub-set (19,250 cases) is used for the validity and the third sub-set (19,250 cases), the previously adjusted models are applied, so that the error measurements calculated using said test set allow estimating the generalization capacity, that is, what performance should be expected for a model over the observations of the studied population. This division was kept in the construction of all 36 models (18 different configurations and 2 predictive techniques).

**Table 5.2** Transformations of the predictor variables

| Predictor variables | Transformation |
|---|---|
| Footprint (FP) | 1/FP |
| Form ratio (FR) | 1/FR |
| Number of storeys (NS) | Exp(NS) |
| Window-to-wall ratio (WWR) | 1/WWR |
| Coefficient of performance (COP) | Log(COP) |
| Energy efficiency ratio (EER) | Log(EER) |
| Heating emission factors (HEF) | – |
| Cooling emission factors (CEF) | – |

**Table 5.3** Matrix of correlations of the transformed predictor variables

|  | FP | FR | NS | WWR | COP | EER | HEF |
|---|---|---|---|---|---|---|---|
| FR | 0.0000 | – | | | | | |
| NS | 0.0000 | 0.0000 | – | | | | |
| WWR | 0.0000 | 0.0000 | 0.0000 | – | | | |
| COP | −0.3283 | −0.0083 | −0.6144 | 0.2251 | – | | |
| EER | 0.0014 | −0.0030 | 0.0000 | 0.0000 | −0.0019 | – | |
| HEF | −0.2482 | −0.0064 | −0.4629 | 0.1694 | 0.7275 | −0.0324 | – |
| CEF | 0.0000 | 0.0003 | 0.0001 | 0.0004 | 0.0013 | −0.0667 | 0.4769 |

On the other hand, the models constructed in this work have been adjusted considering the sets of pairs (predictor variables, dependent variable). In the first setup, both the predictor variables and the dependent variable intervene just like they have been measured; consequently, there are four predictor variables for demand, six for consumption and eight for emissions. In the second setup, four predictor variables have been built, through transformations following the relation they have with the energy demands of the buildings. Specifically, the eight transformations of the predictor variables for this setup, appear in Table 5.2.

Table 5.3 shows the correlations estimated between the transformed predictor variables for the models MLR2, MLR3, PM2 and PM3. It can be seen that there is no presence of severe multicollinearity, that is to say, a correlation between the predictor variables. In this case, there are no correlations with absolute values above 0.5 (without including the constant term).

In the third setup, apart from transforming the predictor variables, the dependent variable is transformed logarithmically, looking for a more symmetrical distribution. Both the Multiple Linear Regression (MLR) models and the Multilayer Perceptron (PM) have been developed over the three setups, so there are six different models which have been adjusted for each one of the dependent variables (Cooling Demand, Heating Demand, Cooling Consumption, Heating Consumption, Cooling Emissions and Heating Emissions).

## 5.4 Comparison with Linear Regressions

Tables 5.4, 5.5 and 5.6 contain three quality indicators of the prediction models. Three measurements have been selected from the diverse existing indicators. The tables also include in the row H information about the size of the hidden layer of each multilayer perceptron.

First of all, in the tables, the $p$-value corresponding to the Ljung-Box test is shown about the possible first order self-correlation in the residue that arises from the adjustment of each model. It would be desirable that the $p$-value of LB would be non-significant, that is, if LB $>0.05$, the null hypothesis that the first order correlation

**Table 5.4** Results for energy demand

|  | MLR1 | MLR2 | MLR3 | PM1 | PM2 | PM3 |
|---|---|---|---|---|---|---|
| Cooling |  |  |  |  |  |  |
| Ljung-Box (p-v.) | 0.145 | 0.447 | 0.439 | 0.609 | 0.101 | 0.378 |
| ECM | 0.497 | 0.552 | 0.552 | 0.032 | 0.117 | 0.072 |
| $R^2$ | 0.357 | 0.206 | 0.205 | 0.997 | 0.964 | 0.986 |
| H |  |  |  | 14 | 12 | 14 |
| Heating |  |  |  |  |  |  |
| Ljung-Box (p-v.) | 0.265 | 0.668 | 0.835 | 0.098 | 0.576 | 0.480 |
| ECM | 0.275 | 0.361 | 0.361 | 0.012 | 0.025 | 0.032 |
| $R^2$ | 0.635 | 0.369 | 0.378 | 0.999 | 0.997 | 0.995 |
| H |  |  |  | 15 | 12 | 14 |

**Table 5.5** Results for energy consumption

|  | MLR1 | MLR2 | MLR3 | PM1 | PM2 | PM3 |
|---|---|---|---|---|---|---|
| Cooling |  |  |  |  |  |  |
| Ljung-Box (p-v.) | 0.717 | 0.918 | 0.930 | 0.649 | 0.232 | 0.398 |
| ECM | 0.226 | 0.159 | 0.128 | 0.013 | 0.014 | 0.019 |
| $R^2$ | 0.953 | 0.977 | 0.985 | 1.000 | 1.000 | 1.000 |
| H |  |  |  | 10 | 14 | 12 |
| Heating |  |  |  |  |  |  |
| Ljung-Box (p-v.) | 0.756 | 0.563 | 0.860 | 0.735 | 0.957 | 0.343 |
| ECM | 0.416 | 0.340 | 0.290 | 0.024 | 0.049 | 0.043 |
| $R^2$ | 0.895 | 0.930 | 0.950 | 1.000 | 0.999 | 0.999 |
| H |  |  |  | 15 | 9 | 11 |

**Table 5.6** Results for $CO_2$ emissions

|  | MLR1 | MLR2 | MLR3 | PM1 | PM2 | PM3 |
|---|---|---|---|---|---|---|
| Cooling |  |  |  |  |  |  |
| Ljung-Box (p-v.) | 0.698 | 0.553 | 0.842 | 0.901 | 0.593 | 0.823 |
| ECM | 0.248 | 0.226 | 0.095 | 0.010 | 0.016 | 0.012 |
| $R^2$ | 0.959 | 0.966 | 0.994 | 1.000 | 1.000 | 1.000 |
| H |  |  |  | 14 | 12 | 15 |
| Heating |  |  |  |  |  |  |
| Ljung-Box (p-v.) | 0.445 | 0.863 | 0.747 | 0.006 | 0.818 | 0.332 |
| ECM | 0.097 | 0.086 | 0.075 | 0.008 | 0.014 | 0.019 |
| $R^2$ | 0.855 | 0.886 | 0.916 | 0.999 | 0.997 | 0.994 |
| H |  |  |  | 14 | 15 | 14 |

of residuals equals 0 could not be rejected. With regard to Ljung-Box, generally speaking results for $p$-values are non-significant.

The second indicator is the Mean Quadratic Error (ECM), that is, the mean of the quadratic residue. The measurement has been calculated on the original scale of the dependent variable, even in the models MLR3 and PM3. This ECM value should be as small as possible.

The third measure, $R^2$, is the linear correlation coefficient to the square between the values observed and the predictions. $R^2$ values range between 0 and 1, with the best values being those that are nearest to 1.

All the models constructed present residues where a significant self-correlation of the first order is not appreciated. While for the ECM and $R^2$ criteria, the model which provides a greater generalization capacity, according to the values obtained in the test set, is the multilayer perceptron adjusted over the original variables, which offers ECM values closer to 0, with an $R^2$ coefficient that is greater than 99% (Tables 5.4, 5.5 and 5.6). The comparison of the MLR and PM models demonstrates a clear superiority to the multilayer perceptron models, substantially improving the quality of the predictions. It is possible to see in Tables 5.4, 5.5 and 5.6 that the PM models do not need to work with the transformed variables, their nature of universal non-linear approximators lets them obtain high quality predictions with the original predictor variables.

Focusing on the demand models, the effect of making transformations is not positively seen in the MLR2 and MLR3 models. As can be seen in Table 5.4, the ECM test increases from 0.497 to 0.552 for cooling and from 0.275 to 0.361 for heating, while the $R^2$ coefficient passes from 35.7 to 20.6% and from 63.5 to 36.9%. However, the models based on the multilayer perceptron have not needed to work with transformations of the predictor variables. The PM1 model provides a lower ECM test than those of the PM2 and PM3 models, with $R^2$ coefficients of 99.7% for cooling and 99.9% for heating. This shows the non-linear nature of the multilayer perceptron,

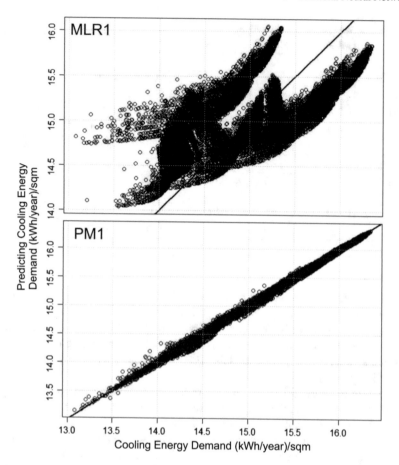

**Fig. 5.3** MLR1 and PM1 test, cooling energy demand (kWh/year)/m$^2$

capable of providing a good fit when the proportion between the predictors and the dependent variable is not of a linear nature. This better behaviour seen for the models based on the multilayer perceptron is even cleared in the point clouds formed by the pairs (dependent variable, prediction) (Figs. 5.3 and 5.4).

As an illustration, Figs. 5.3 and 5.4 show the points clouds for the best linear regression model for demands (MLR1) and the best multilayer perceptron model (PM1), for the cooling and heating variables in the test set. The fit provided by PM1 is clear, graphically showing the high quality of the model obtained. It is seen that the points cloud for the linear regression models has a much greater spread than the points clouds for the PM models.

In regards to the consumption models, the effect of making transformations both of the predictor variables and of the dependent variable is seen positively in the MLR2 and MLR3 models. The ECM test is reduced between the MLR1 and MLR3

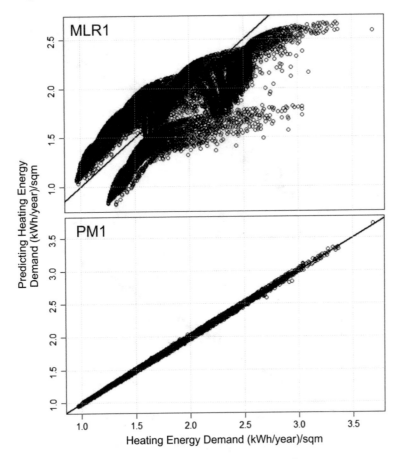

**Fig. 5.4** MLR1 and PM1 test, heating energy demand $(kWh/year)/m^2$

models from 0.226 to 0.128 for cooling and from 0.416 to 0.290 for heating, while the $R^2$ coefficient passes from 95.3 to 98.5% and from 89.5 to 95.0% (Table 5.5). Just like what would happen in the models based on the multilayer perceptron for the demand, it has not been necessary to generate transformations of the predictor variables, obtaining an ECM of 0.013 with an $R^2$ of 100.0% for cooling and a heating model with an EMC of 0.024 with an $R^2$ of 100.0%. For that reason, in Figs. 5.5 and 5.6, it can be seen that the multilayer perceptron continues to provide a better fit between the non-linear dependent variable, obtaining a lower spread even in the heating consumption variable which has a greater difficulty of fit in the regression models in spite of the transformations both of the predictor variables and the dependent variable.

The emissions models in spite of being formed by eight predictor variables have similar behaviours to those of consumption, obtaining the best result in the regression models, case where the predictor variables and the dependent variable have been

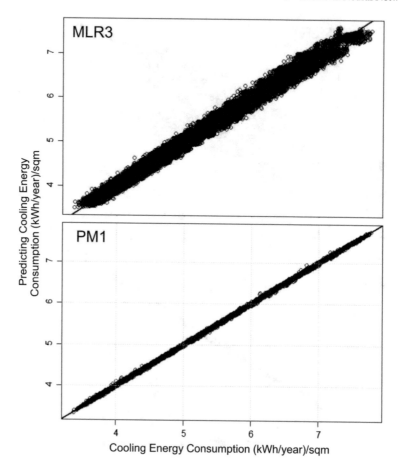

**Fig. 5.5** MLR3 and PM1, test, cooling energy consumption (kWh/year)/m²

transformed. The MLR3 model reduces the ECM for cooling from 0.248 to 0.095 and from 0.097 to 0.075 for heating, with $R^2$ passing from 95.9 to 99.4% and from 85.5 to 91.6%. Just as occurs in the previous cases, the multilayer perceptron model has a good fit both for the untransformed and transformed variables, finding the best fit in the PM1 model with an ECM of 0.010 for cooling emissions and of 0.008 for heating emissions, with the $R^2$ for the first emissions model sitting at 100.0% and for the second at 99.9% (Table 5.6). It is necessary to indicate that the transformations of the variables in the PM models do not only worsen their performance but their ECM too and in some cases their $R^2$ as well (Tables 5.4, 5.5 and 5.6).

As an illustration, Figs. 5.7 and 5.8 show the points clouds for the best linear regression model for emissions (MLR3) and the best multilayer perceptron model (PMI) for the cooling and heating emissions variables in the test set. The fit provided for PM1 is clear, graphically showing the high quality of the model obtained.

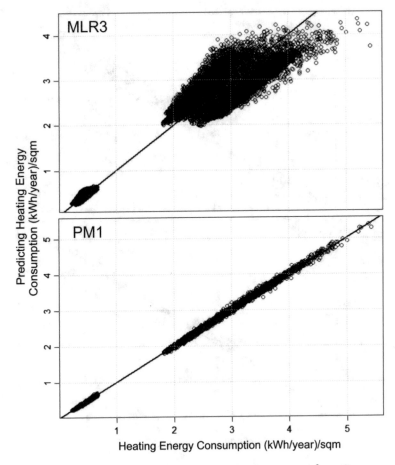

**Fig. 5.6** MLR3 and PM1, test, heating energy consumption (kWh/year)/m$^2$

Therefore, the results show a better performance of the models based on the multilayer perceptron, not only for the analysis of the numeric indicators, but by the graphical representation on comparing observed values and predictions, where an almost perfect fit is shown for the neural network model built on the first setup, that is, considering the four, six and eight predictor variables. The PM1 models produce the best results working with ECM and $R^2$, and for each dependent variable, the points cloud of the model based on the multilayer perceptron shows a stronger linear relation in comparison with the corresponding linear regression model. The improvement of the fit stands out especially in the predictive models for cooling and heating demand. It is true though, that the MLR models do not always improve with the proposed transformations.

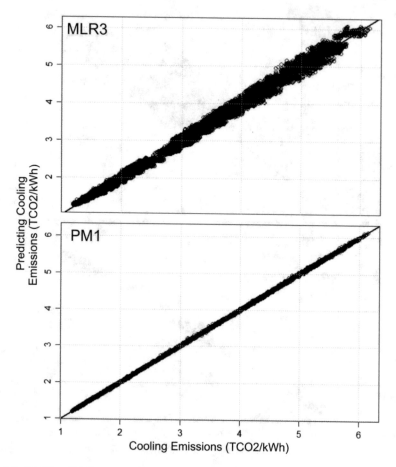

**Fig. 5.7** MLR3 and PM1, test, cooling emissions (TCO$_2$/kWh)

## 5.5  Discussions

A procedure to develop training data set for ANN following a quasi-static method calculation procedure such as ISO 13790:2008 has been successfully implemented, generating a database of 77,000 cases, being statistically representative of a concrete building typology, in this case, office buildings. The authors consider that this fact remains particularly important because one of the main concerns when using ANN is the amount of data that can be used as the training set. This procedure, which relies on a verified methodology, allows for the generation of considerable amounts of data to be used not only in the present but also in further studies.

The statistical models that best reproduce the results of the ISO 13790:2008 standard, are those developed from the multilayer perceptron, obtaining an $R^2$ of 99.8% in cooling and 99.9% in heating with ECM values of 0.0008 and 0.0001

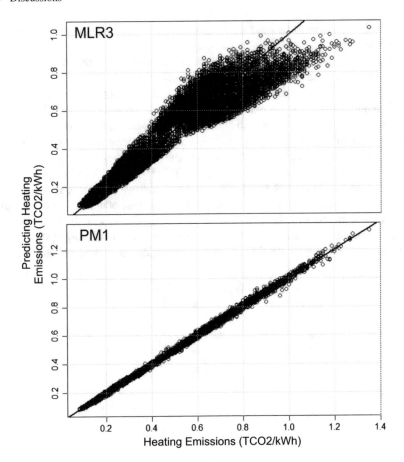

**Fig. 5.8**  MLR3 and PM1, test, heating emissions ($TCO_2$/kWh)

without needing to transform the dependent variable or the predictors. The linear regression models, however, obtain higher performance when the predictor variables are transformed, obtaining $R^2$ values closer to 85% both in heating and in cooling, and ECM values between 0.0269 and 0.0606. Through this, it can be confirmed that the models created starting from the Artificial Neural Network (ANN) have a greater precision to substitute the calculation procedures established in the norm when the constructive standards and the internal loads are suitably defined.

An ANN model has been obtained for the prediction of multiple variables regarding the energy demand, energy consumption and $CO_2$ emissions for a statistically representative set of data regarding office buildings in Santiago. The ANN model has shown a satisfactory performance even when dealing with variables without transformation, obtaining values of $R^2$ over 0.994 for all models and non-significant p-values. It has been determined that, so there is no collinearity between the predictor variables, these must be transformed, mainly IR which is linked to the building's volume and, therefore, the number of storeys.

It is considered that the results from this research can open roads towards the future implementation of advanced calculation methods regarding energy demand, energy consumption and $CO_2$ emissions, being the first research to bring these methods into the Chilean legislative framework. ANN models can provide simplified and faster calculation methods that are able to enhance the implementation of ISO 13790:2008 into the Chilean context. Designers will be benefitted as they will be able to estimate the depletion of resources for their projects in the early stages of design; FP, NS and FR define the basic shape of the construction, WWR deals with the amount of glass in the façade and finally IR is related to the constructive standards intended to be implemented. Additionally, as this country is in the process of updating the legislative framework with regard to building energy efficiency, the values obtained from this study can be of use when developing future standards, establishing realistic goals that can be effectively accomplished by developers, designers and stakeholders.